THE ULTIMATE GUIDE TO
URBAN FARMING

THE ULTIMATE GUIDE TO

URBAN FARMING

SUSTAINABLE LIVING IN YOUR HOME, COMMUNITY, AND BUSINESS

NICOLE FAIRES

Skyhorse Publishing

Skyhorse Publishing books may be purchased in bulk at special discounts for sales promotion, corporate gifts, fund-raising, or educational purposes. Special editions can also be created to specifications. For details, contact the Special Sales Department, Skyhorse Publishing, 307 West 36th Street, 11th Floor, New York, NY 10018 or info@skyhorsepublishing.com.

Skyhorse® and Skyhorse Publishing® are registered trademarks of Skyhorse Publishing, Inc.®, a Delaware corporation.

Visit our website at www.skyhorsepublishing.com.

10 9 8 7 6 5 4 3 2 1

Library of Congress Cataloging-in-Publication Data is available on file.

Cover design by Rain Saukas
Cover photo credits: John Faires, Shutterstock, and iStockphoto
Interior Photograph credits: Nicole Faires, Jared Price-Morin, John Faires

ISBN: 978-1-5107-0392-6

Ebook ISBN: 978-1-5107-0393-3

Printed in China

For Ana, Autumn, Rainn, and Rhylan . . . and especially for Jared.

Contents

Introduction

My first book was about homesteading, but not the modern-day hobby homesteading of yuppie housewives. My homesteading guide was about returning to the lost skills of the 1800s in case of a total social apocalypse. Zombie attack? No problem. Someone dropped you in the woods with just a pocketknife? Piece of cake.

As a child, I was an avid reader of children's books like *Little House on the Prairie* and *My Side of the Mountain*. Who hasn't dreamed of living in the woods? I planned to buy some land and retire to my self-sufficient paradise. I learned the hard way that land is very expensive, and if you live in a rural area it is even more difficult. There just aren't many jobs in the country and they don't pay well. Growing up in Montana, I quickly learned that there are generally two types of people in the woods: people living below the poverty line so that they can be free amongst nature, and rich folks who like the view from their million-dollar summer lake homes.

The children of country people don't usually stay in the country. Eighty percent of Americans and Canadians live in the city, because that's where the jobs are. In many ways, it's less stressful to live in the city. The rents are high, the cities are crowded, and day-to-day necessities are pricier. However, the grocery stores are nearby, mass public transportation is convenient, and the schools are often superior. Therefore, as farmers retire, their kids aren't taking over their jobs because it would mean forfeiting the comfort and security of the suburbs or city.

So, although I spent most of my childhood on a rural hobby farm in Montana, I have lived in numerous suburbs and cities during my adult life, including in Phoenix, Las Vegas, Nevada, Vancouver, and Washington (which is really a suburb of Portland). I also experienced small-town life: from the desert beauty of Apache Junction, Arizona and bleak Elko, Nevada to blue-collar Columbia Falls, Montana. Now I live on Vancouver Island in the elegant capital city of Victoria. Eventually, I would like to settle down in the country as I've always dreamed, but that doesn't mean the city can't be a great place for farming. The city is an important part of the future of food; people have to eat, and that's where it happens most.

Each city has its own food culture and flair. There's always some hole-in-the-wall little restaurant with the best eggs Benedict, or with pie that's better than your previous experience. Even though people have to eat, they love to eat local fare that has its own cultural identity.

We have access to some of the best food in the world, and yet we struggle with heart disease and obesity rates directly related to our diet of fast food or factory-made foods. Even the 20 percent who live in poverty-stricken rural areas have limited access to fresh food. This is due to the plight of small farms, which I researched for my book *Food Confidential*. Small farms are in danger not only for economic reasons, but also because corporations are in bed with the government to regulate them out of business. For example, industrial chicken producers work their hardest to make it impossible for small farmers to sell homegrown meat, and food conglomerates spend millions of dollars to lobby for more relaxed organic standards in order to make it easier for them to sell a high-profit product labelled *organic*.

Our issues with food will only get worse with climate change. The recent drought in California is just a sign of things to come. Urban agriculture is the solution to our problems. It is the best, cheapest, most reliable way to bring the freshest food to the most people. This book is about how to implement that solution.

There are tons of books available nowadays that call themselves "urban homesteading" or "urban farming," but all they do is teach a fancier version of backyard gardening. There is a distinction between farming and gardening, and frankly, it's a bit of an insult to farmers when you call a small vegetable garden a farm. This book is intended to show the difference between farming and gardening, and also how you *can* be a farmer, even in an apartment. This requires some hard work, but it is possible. It shows you how you can grow your own food in your suburban backyard as well as how you can become a commercial grower in any city.

Warning: Urban agriculture isn't legal in many places. *Food Confidential* talks a lot more about this and the struggles that backyard farmers should consider. To find out the legalities of urban farming in your area, you will need to learn more about your municipal bylaws and find your local food security organization.

Urban farming is not only a solution for our future as humans face climate change, corruption, and agricultural problems. It is also a solution for families who need access to fresh vegetables and fruit without the burden of shame and helplessness that food banks and care packages come with. Growing food provides a tremendous source of pride, self-reliance, and accomplishment. It brings communities together. It solves hunger and health problems. It can also provide income. There is no other vocation that can do all of this so holistically.

The Ultimate Guide to Urban Farming is a comprehensive system for creating and building urban farming as a major source of fresh foods in a city within a relatively short period of time. You can do it right now. Let's get started!

1 | Community & Business Planning

⌃ Farming can be a family business. My daughter enjoyed gathering pill bugs.

DEFINING URBAN AGRICULTURE

"All cities are mad: but the madness is gallant. All cities are beautiful, but the beauty is grim."

—*Christopher Morley*

The Food Distribution System

The first grocery stores in North America were general stores and only rarely offered fresh food. Instead they sold things that people couldn't grow or make themselves and things that could be stored away for months at a time, such as canned goods, flour, and sugar. When you went to the general store, you brought your goods to trade or your little bit of money, walked directly up to the counter, and told the owner or clerk exactly what you wanted. He would tell you the price, you haggled over it, and then you handed over your beaver pelts.

Then came Piggly Wiggly in 1916, which changed everything. Clarence Saunders had the novel idea that shoppers could just take things off the shelf for themselves. It was called the amazing "self-serving store," but he didn't stop there. The real game changer was his newfound ability to buy in high volume at lower prices. He became one of the very first wholesale buyers. No longer would there be haggling or time-consuming debates with customers over prices— he already offered the lowest price. Grocery store chains began popping up everywhere in the 1920s, and soon there were tens of thousands of them across America. By 1940, all these little stores had conglomerated and started creating *supermarkets*, with butchers, bakers, produce managers, and dry goods all under one roof. By the 1950s, these supermarkets had already migrated to the suburbs, leaving the little corner groceries in the dust. It only took thirty years before the general store had disappeared.

In the process of migrating from the small store to the supermarket concept, the way that food moved around the country changed as well. Previously, almost all farmers had sold directly to the customer or to a produce man who sold directly to customers. There was never more than one middleman. But when grocery stores became larger and larger and more distance was created between the farmer and the customer, food distribution became much more complex.

Here's how food distribution works today: The farmer sells to food wholesalers through a broker. The broker negotiates a deal between the wholesaler and farmer, taking a cut in the process. The wholesaler then sells the food at a marked-up price to the grocery chain in large quantities. This food is then trucked to a warehouse where it sits for a week until it can be placed on a shelf in a store.

Then there are *marketing boards*. In the United States, the boards are organizations that farmers can voluntarily belong to and act as policy watchdogs and advertising representatives. Although they really only represent factory farms, they serve the interests of farmers who pay a rate to the board based on how much they produce. The farmers make the choice to pay this fee because it provides insurance that they will continue to receive government subsidies and retain control of their distribution system. In Canada, however, these boards are much more insidious. If the farmer produces a "regulated" product, he must then comply with whatever policy the board decides. These Canadian boards have the ultimate power to fix prices, require farmers to sell through certain distributors rather than directly to customers, and help manage subsidies.

To sell a regulated product on more than a backyard scale, the farmer must purchase a quota on a unit price. Today, there are no quotas left because they were bought up years ago and they are inheritable. The boards don't create more quotas. These boards have completely locked up production of most food staples in Canada, including cabbage, potatoes, carrots, milk, greenhouse tomatoes, wheat, and more. It is impossible for young farmers to break in unless they are directly related to an aging farmer who owns a quota. If someone chooses to grow and sell a regulated product without a quota, these boards have the right to seize their property, levy ridiculous fines, and even send them to jail.

Now add to this the lobbyists and government agricultural departments working to manipulate prices, plan subsidies, and create multimillion-dollar marketing campaigns aimed at consumers. All of these actors need to be paid, and the people at both ends of this chain are the ones who pay for it: the farmer and the consumer.

Food prices rose drastically as the recession began to build momentum in 2011. We all felt it. People assume this is because of "the economy," or climate change and bad weather, but this is only partly true. Raw goods in their most basic form are called *commodities*, and when people hear that word they usually think of things like copper or coal. But one major commodity you don't hear much about is food. During the last decade, investment banks like Goldman Sachs had begun helping investors *speculate* on food crop. When speculators bet on *futures* of corn and soybeans, they pay a price for a future crop of food and hope that in the end it sells for more than they paid. Previously, futures trading was a way of stabilizing the market. Farmers would make a deal to sell future crops for a specific price, which gave them some insurance and predictability. However, when this system was deregulated, the speculators were able to cause problems by manipulating it in ways that drove up prices instead of stabilizing them.

Here we have this global food distribution system, which trades food in the same way as any other commodity, is very dependent on petroleum to produce it and move it around, and is very susceptible to changing markets and economies. Drought, oil prices, and even social unrest can all make it hard for people to afford a basic necessity like food.

This is where urban agriculture comes in and why we need to look at Cuba.

Havana

Havana, Cuba has only recently been recognized as a model of the city of the future, not because it has some kind of futuristic cityscape (far from it), but because:

- 60 percent of the total food supply is produced within the city limits;
- 90 percent of the fresh vegetables and fruit are grown locally on an estimated eighty-five thousand acres of land (8 percent of the total land), versus 90 percent of the supply of vegetables/fruit imported into the US and Canada;
- Havana's population is more than 2 million (making this a huge accomplishment);
- this was created due to a post-oil era: without transportation, food has to be local;
- Cuba has 6 percent food insecurity versus US at 14 percent (FAO, Coleman-Jensen);

- every fifteen houses have a growing space; and
- agricultural centers are located in every neighborhood.

Organopónicos are urban farms in Cuba that run without petroleum. They were created during what Cubans call the Special Period (*Período especial*), which began in 1989 when the Soviet Union was dissolved. Cuba found itself in a very sudden and severe economic depression based mostly on the fact that their trade deals with the Soviet Union were gone, including petroleum. Every industry that relies on petroleum was hit, including agriculture. As soon as this happened, a huge national effort began to build up a local food supply that did not rely on oil; the alternative was mass starvation.

As other industries in Cuba collapsed, the government encouraged the unemployed to work in farming. Cuba has a high percentage of highly educated people because university tuition is free. These scientists, engineers, and doctors used their education and ingenuity to create new low-cost solutions in organic agriculture. That said, a lot of people starved anyway.

▲ Balcony broccoli and intensive container production.

The solution started a little late and it took a while to catch up, which brings us now to our own definition of urban farming.

Urban Farming Defined

Urban farming should really be used only to describe intensive food production in or near a city. Although there are many wonderful gardens designed to look great and grow food, often called "edible landscaping," farming is an activity that grows as much food in a space as possible, and this is where it deviates from urban homesteading. An urban homesteader is dabbling in a wide variety of self-reliant skills to raise food and make things for their own family, like knitting, canning, and beekeeping. A farmer, on the other hand, is focused on producing food, and that's it. A farmer's goal is to produce a lot more food than she can eat herself. This does not mean that the farmer *doesn't* do a lot of other stuff. But a farmer is putting a lot of energy into a growing system for food production.

Not only is Havana the best model for changing our food distribution system because it really works, it is also a great example of why we should make such a change. A sudden collapse of just one of the systems that keeps our food distribution in place leads to hunger very quickly. If we want real change, we have to move from being hobbyists to serious farmers who act as if our lives depend on it. The food insecurity in the United States right now is as high as it was in Cuba during the Special Period, so our lives actually do depend on it.

PERMACULTURE

"[Permaculture] is the harmonious integration of the landscape, people, and appropriate technologies, providing food, shelter, energy and other material and non-material needs in a sustainable way."

—*Bill Mollison*

Today, definitions of permaculture differ. It has been described as a way of life. A culture. A philosophy. At its very core, permaculture is a way of designing all human systems so that they integrate harmoniously with ecology. It grew to include community systems, cultural

▲ An herb spiral, which represents some of the basics of permaculture: a functional, connected growing system.

ideologies, business, art—every facet of human life.

What had originally started out as *permanent agriculture* ended up meaning *permanent culture* because the idea encompassed much more than just agriculture. Author and scientist Bill Mollison described it this way: "Permaculture is a philosophy of working with, rather than against nature; of protracted and thoughtful observation rather than protracted & thoughtless action, of looking at systems in all their functions rather than asking only one yield of them." One great permaculture teacher, Toby Hemenway, described it even more simply: "Turn every liability into an asset."

When Cuba faced its economic crises, the country was fortunate that a small army of permaculture teachers arrived from Australia to share what they knew about creating an efficient, low-cost food-producing system. This became the basis of the agriculture that exists in Cuba today. Another remarkable part of this is the idea of information sharing: Cuban farmers shared everything they knew, which has created a level of standardization. Farmer-

to-farmer knowledge sharing is a part of the support system we will talk about shortly. This book focuses a lot on permaculture principles as a proven system for creating real change and sustainability.

Permaculture Ethics

Ethics guide our behavior and are the vehicles by which our destiny manifests itself. There are three ethics of permaculture, and they are fairly simple:

Care for the Earth: All things, living or nonliving, have intrinsic worth.

Care for People: Humanity is cared for through self-reliance and community responsibility.

Give Away the Surplus: The surplus should be shared to fulfill the other two ethics.

In a nonpermaculture system, everything is used once or twice and then thrown away into the water and air, never to be seen again. This is a *linear* system because everything makes a straight line from the source to the landfill. Sustainable systems, by contrast, are circular. The used items go back to their source, where they can go through the natural recycling process of the earth and be used again, using very little energy. The same is true of permaculture ethics. When each resource or living creature is valued rather than exploited or destroyed, and people care for themselves as well as their community, an excess of resources is the natural result—and these can then be used to care for the earth and people again.

Principles of Permaculture

Permaculture may be a very creative and imaginative method of design, and it may work with some highly variable pieces, but it still follows some basic principles. Different permaculture groups may phrase these principles differently, but the meaning is the same. Most will include twelve principles or more, but I have combined some of these together for simplicity.

1. Every thing is connected to and supported by everything else.
2. Every thing, or *element*, should serve many functions. Students of design usually learn to make things look nice while being functional, but permaculture focuses on function alone.
3. Functional design is sustainable, and it provides a useful product or surplus. If it doesn't, it creates pollution and work. Pollution is an overabundance of a resource, or something that is simply not used. Work results when one element doesn't help another element.
4. Permaculture maximizes the useful energy in any system (or, put another way, decreases the waste of energy).
5. Successful design serves the needs of people and provides many useful connections between elements, or *diversity*.
6. If there is pollution, then the system goes into chaos.
7. Societies, systems, and human lives are wasted in disorder and opposition.

To stop this vicious cycle, we only use what we can return to the soil, and we build harmony (cooperation) into the functional organization of a system.

THE SUPPORT SYSTEM

"There are two spiritual dangers in not owning a farm. One is the danger of supposing that breakfast comes from the grocery, and the other that heat comes from the furnace."
—*Aldo Leopold,* A Sand County Almanac

I believe the basis for Cuba's success in transitioning to a localized food system with so much of the population involved in urban agriculture comes down to the government-created support system. Living in a postpetroleum age impacted the availability of fertilizers and pesticides, forcing them to use organic production. Information became key. This information and support came in the form of horticultural clubs for every neighborhood as well as agricultural organizations (both governmental and community-based) with the goal that "no space in the city should go uncultivated." These groups and centralized locations offer the following:

The Ultimate Guide to Urban Farming

- data and information about growing in the local area
- shared equipment
- travelling vets and animal registries
- farmstands and markets in every neighborhood
- free seeds, local seed development, and exchanges
- workshops and collaborations
- seedling greenhouses
- free compost from organic waste collection
- organic pest control
- agroecological libraries
- regional competitive ranking systems
- microloans

Schools often act as centers, from kindergartens to universities, but in the absence of an available location, a shipping container is placed to become an agricultural hub. In the United States and Canada, we have seen very few cities embracing urban agriculture on a municipal level, and as far as I can find, there are very few organizations that seek to create a neighborhood-level support system for urban farmers. This kind of support was created in Cuba out of necessity because of the lack of gas for cars, but the end result would be the same anywhere: it is accessible to everyone regardless of income, it is convenient, and it brings the community together. Agricultural extensions exist to help farmers in rural areas, but they really should be expanded to include the urban environment.

Creating Agricultural Hubs

In an effort to research a system for North America that would follow Havana's model of neighborhood-based agricultural hubs, I began creating an inventory of available resources for my own neighborhood in Victoria, British Columbia based on the list of supports Cuban farmers have available to them. I then divided my city up into its natural neighborhoods. Cities often divide themselves into communities based on history, geography, and culture; for example, my city is made up of regional districts like Fernwood and Saanich, which operate community centers and independent police departments. It also has within it First Nations territories; Chinatown; a mosque that acts as a focal point of a local community; and universities, which are their own microcosms of culture. Each of these layers has its own needs, resources, and challenges in obtaining local food.

I picked the minds of people in my area who had experience with both food hubs and urban farmers. Many urban farmers are young, new farmers and a large number are women. I also spoke to farmers, city councils, agricultural land representatives, young farmers' coalitions, municipal policy activists, and consumers. All of these players were in absolute agreement that the food distribution system needs to change on a fundamental level. They agreed that there needs to be a cultural shift in government from agribusiness to a small, local, sustainable model. They also agreed that there are probably too many organizations working toward the same goals but not communicating well with each other or using resources efficiently.

People are not in agreement about how to change things, although there are some generally accepted ideas.

The people working in policy and government to promote local food are wildly supportive of standardized solutions using cheap technology. Databases like a wiki I developed to keep track of my neighborhoods were very interesting to these groups because they represent a simple solution to sharing information, mapping data, and quantifying results. This is important in government because funding is largely based on simple ideas that can easily be proven with numbers.

The food hub organizations, such as community centers, food banks, and student-led activist groups, preferred to create small, short-term projects focused on the needs of an individual, microregional group of people. For example, a food bank would want to start a farm to grow food, using a student-led organization for volunteer support. This type of project was attractive to them because it was easily fundable. However, it was also not sustainable, because it was run on grants, and it was short-lived, because the people running food banks are likely volunteers, not farmers.

Farmers have found that many people want to support local food, but the group of people who are buying locally is very small. Farmers also can't support more local farmers' markets because they can't physically be at most locations. In addition, they would need to be better at marketing and business, but they don't have the time to do so because small, sustainable farms are labor-intensive. They would like a better distribution system, training programs, and more funding for their farms.

In the end, the ability to organize neighborhood-based support for urban farming depends on a strong relationship between the government, local food organizations, and farmers. There needs to be an active government advocate, such as a city councilor, and key community

activists who can support the farmer-to-consumer connection. The Cuban model allows free land, free supplies, and free education to anyone who is interested in growing food. It is imperfect and filled with problems, but it is a system that is producing great outcomes.

Farmer-to-Farmer Information Sharing

There are two groups of farmers: protective farmers and open-source farmers. Protective farmers tend to hoard their growing information as business secrets. Open-source farmers share all of their production information freely with other farmers.

Does sharing business information hurt your farm? In most industries, trade secrets should be closely guarded, but farming is not a typical industry. The most important reason for this is that demand is so incredibly high for local food, yet farmers don't realize that they aren't competing against each other. They are competing against the grocery stores, not other farmers.

COMMUNITY

"The community I desire is not grudging; it is exuberant, joyful, grounded in affection, pleasure, and mutual aid. Such a community arises not from duty or money but from the free interchange of people who share a place, share work and food, sorrows and hope. Taking part in the common life means dwelling in a web of relationships, the many threads tugging at you while also holding you upright."
—Scott Russell Sanders, "The Common Life" in Writing from the Center

What Community Is (and Isn't)

Farming is a group-oriented activity. Therefore, historically, permaculture has been heavily focused on the building of *community*. Unfortunately, it can be difficult to use land in an urban environment for farming activities without falling victim to tons of expenses or annoyed neighbors. To this end, a strong community must be built up. The community should consist of like-minded people who can pool their resources to acquire land or create protection around the activities of a local farmer.

A diligent community is able to apply the third permaculture ethic: share the surplus. Individuals will share with their small community of friends, who will share with the greater community of not-so-like-minded neighbors. That is when real social change begins.

To summarize, permaculture applied to community has three guiding principles:

- Create a support system and teaching network for like-minded people to be able to accomplish their goals.
- The only thing this community must have in common is the belief that humanity should live sustainably and self-reliantly.
- Self-reliance does not come from independence from people, but from independence from business and organizations that rely on products and resources from faraway places.

Starting a Community

A good starting point in developing community is to join an existing one. Community develops naturally, so it is just a matter of finding like-minded people within that group of people who have similar goals.

The quickest introduction to that community is often made in a food hub. A food hub is a regional business or organization that actively manages the aggregation, distribution, and marketing of source-identified food products primarily from local and regional producers to strengthen their ability to satisfy wholesale, retail, and institutional demand.

On a small scale, a food hub would be a communal kitchen in a community center that offers a meal preparation and sharing day, a church that has a weekly breakfast, or a farmers' market or café that serves locally produced food.

Community & Business Planning

You are likely to find a food hub within any of these organizations:

- community center
- elementary school
- parent advisory council (or parent teacher association)
- farmers' market
- church
- food bank
- local food and farm initiatives

DECISION-MAKING IN COMMUNITY

"A whole stream of events issues from the decision, raising in one's favor all manner of unforeseen incidents and meetings and material assistance, which no man could have dreamed would have come his way. Whatever you can do, or dream you can do, begin it. Boldness has genius, power, and magic in it. Begin it now."

—*W. H. Murray,* The Scottish Himalayan Expedition

Frequent communication is important, even in the smallest group. Everyone needs to be on the same page, and any decision that impacts everyone else should be made in consensus. This is a hassle and incredibly challenging, but it's necessary. There comes a point during a community's evolution where one person feels left out of the process. One discontented person can cause the collapse of the entire process.

The group must be moving steadily forward toward a preplanned goal in order to continue as a community. Otherwise, people naturally lose motivation or run out of energy.

One effective method of making communication more efficient is in keeping the groups small. The first core group of people is often brought together by a common purpose, but as soon as more people join, things get complicated. All of those extra people have their own reasons for being there. These people should be formed into small groups according to their individual personalities and interests, so that each has something smaller to manage. Each group should have a spokesperson and a secretary. The secretary keeps meticulous records of everything said and done, and the spokesperson reports to the greater community. This may seem very formal and traditional, but this "committee" structure has proven itself to be functional even when individuals aren't very experienced with it or have personalities that are difficult to work with.

Not everyone gets along, and no one communicates effectively all the time. Creating an environment that is honest about this basic fact, and works around it, is the key to success. Flexibility is part of this. When a task is done, people can move on to something else. When two people aren't working well together, rather than taking it personally, recognize that sometimes personalities clash and move on. One of the greatest tools in helping people learn to do this is having a common mission. A mission is more than a goal—it embodies the values and direction the group is moving in. This must be decided as a community and written down, and everyone should be reminded of it frequently. Every decision made by the community should be held up to the mission statement to make sure the group is sticking to its values.

Dissent is expected. Consensus is difficult, because if someone disagrees with the group, the whole group cannot move forward. However, there are very few decisions in permaculture that must be made immediately. A single individual may hang up a decision for quite a while, but that is one of the drawbacks of community that must simply be tolerated. It is very important that the founders of the community have communication and consensus decision-making training so that they can facilitate the communications of the group.

The other greatest tool in community is food. Just because you are getting together for a business meeting to discuss difficult topics doesn't mean you can't do it over an amazing potluck dinner of homegrown food.

BUSINESS STRUCTURES

"The good health of a farm depends on the farmer's mind. The good health of his mind has its dependence, and its proof, in physical work. The good farmer's mind and his body—his management and his labor—work together as intimately as his heart and his lungs. And the capital of a well-farmed farm by definition includes the farmer, mind and body both. Farmer and farm are one thing, an organism."

—*Wendell Berry,* The Gift of Good Land

Sole Proprietorship versus Partnership

The usual business model for a small business is sole proprietorship. Every business owner that I know, including myself, who has also been part of a

business partnership has watched it fall apart due to management differences. This is especially dangerous when working with friends and family.

States and provinces vary as to business requirements, but there are major tax benefits to registering your business as a farm. Farms usually have special status and deductions.

Starting a Farm Co-op

A cooperative (or co-op) is a collaboration between like-minded individuals who are working toward a common goal. This is a good alternative to a partnership if there is someone you'd like to work with. Cooperatives are increasingly recognized for their ability to make good things happen—so much so that the United Nations partnered with the International Co-operative Alliance to declare 2012 the International Year of Cooperatives. Cooperative enterprises work differently from other businesses in that they are based on needs and values rather than profits. There are many different forms of cooperatives, but what we are most interested in is agricultural co-ops. Cooperatives aren't just little groups run by Birkenstock-wearing health nuts. Companies like Ocean Spray, Ace Hardware, Sunkist, and the Associated Press are all cooperatives.

A farm co-op has shared ownership and a larger labor pool, which produces more crops. When production increases and is pooled together, it does two things: there is more likely to be a good crop, and it lowers the costs of production. Farmers' co-ops can either be formed of several privately owned farms, or something increasingly common, a farm owned cooperatively. Forming a farm this way is the most viable way for young people to acquire large acreage and become financially successful without too much individual burden of debt. It is not, however, a commune. It's strictly a business arrangement. Historically, these farms do not benefit the poorest people in the co-op because it takes a significant financial investment to start a good farm, but once the strongest members begin it, opportunity exists to allow others in with smaller investments.

Urban farm co-ops usually do not own the land outright because the cost in the city is almost always prohibitively expensive. Farming could not ever pay off the mortgage. Instead, the co-op acts as an investment pool for leaseholding, equipment, and inputs (fertilizers, seeds, and other yearly expenses). However, even though the farmers are sharing quite a bit, it is also important that each farmer-member have the ability to have his or her own land space and crops. Farmers' co-ops often have a slightly different voting strategy based on production. Because some farmers produce more crops than others, they carry more of the risk and therefore deserve a more powerful vote. Votes may be added based on tons of food or the economic value of crops.

There can't be any outside investment when starting a farm cooperative for the purchase of land or equipment, or for selling crops. This is one of the basic principles of co-ops, because doing so creates a conflict of interest. Rather than focusing on building infrastructure or growing crops, the secondary goal of an interested third party can be distracting or, worse, destructive. To start a farmers' co-op, the first investment goal is for acquiring and developing land. This may be urban land, either borrowed or leased, or rural land, which can be bought or leased. Whether the land is free or not, money must be available to build the land up for farming in the form of inputs, raised beds, high tunnels, irrigation, and fencing. This part of the investment should be pooled, with that same amount saved in a fund that can buy out farmers who choose to leave the cooperative. In this situation, the land would not be owned by any one person, but rather owned by the co-op, and members would own shares based on their investment.

For those interested in this kind of serious involvement in farming, a few key pieces of information are needed. This knowledge is not always freely available, and yet it is extremely important for people to know when getting into agriculture.

A farm in a rural area requires an investment of at least $200,000, depending on the price of the land and what kind of infrastructure is on it. At least $60,000 of that has to go to developing the growing area, inputs, and tools. A half-acre space with a couple of homemade high tunnels costs at least $15,000 to prepare for intensive growing, and for a co-op, two acres is the minimum for a profitable enterprise. It may be possible to find some land for $140,000, but keep in mind that it must be within the 100-mile radius of a city, and preferably within 45 miles at the most. That's the most realistic maximum driving time for delivering produce, so the closer you are to your market the better.

A project in an urban area using borrowed land (such as SPIN farming, or Small Plot INtensive farming) requires an investment of at least $30,000. This is used to purchase inputs, a rototiller, refrigeration equipment, a truck, and other tools. These can be shared among several urban farmers operating out of backyards or leased land. The profits per square foot are higher in the city because the initial investment is so much lower and the market price is higher, but these farms tend to be smaller. The smaller the farm, the less likely the farmer can make it work without a second job to support it. An equipment (and labor) co-op can make it possible for urban farms to do things on a larger scale. For example, many use old refrigerators, but a professional walk-in cooler can hold an entire crop. This is so important when harvesting for a Saturday farmers' market that opens at eight in the morning, because everything needs to be harvested by Friday.

A second kind of farmers' cooperative manages only product marketing for a group of farmers. Many established farms today are members of a cooperative, which they formed to purchase their own product. The co-op, as a separate entity, purchases the crops of all the farmers and then handles the selling and/or distribution. Usually that means handling negotiations with a wholesaler, but today a co-op can become a direct selling machine, whether in the form of a CSA (Community Supported Agriculture farm, which sells shares to members) or a physical market. The labor involved in distributing farm product is taken from the farmers so they can spend more time growing, and there is much less risk because if one farmer has a shortfall, the other farms can fill it. The cooperative can also purchase expensive equipment, like the large coolers and refrigerated trucks that a single farm might not be able to afford. It is important to remember that although a marketing cooperative like this is for making money, the ultimate goal is not higher profits. In fact, profits might be a little lower for the

▾ A weekly delivery from my CSA.

farmer on a per-item basis because he is selling at a lower-than-retail rate to the co-op. The benefits are in cost and labor savings, shared risk, and less stress.

Marketing co-ops need someone in charge with a strong marketing background. This means understanding branding, current Internet selling techniques like social media, and the ability to keep track of current food events and policy. Coordinating the harvest of several entire farms, handling proper processing and storage for optimum freshness, and managing multiple types of customers takes a skilled person. Marketing co-ops will usually have enough products to run a CSA, show up at the farmers' market, sell to restaurants, and possibly run a full-time farm stand. This manager will need several assistants.

The CSA—Community Supported Agriculture

Don't start out financing your first farm equipment by creating a CSA. Community Supported Agriculture operates on the principle of shared risk, but customers expect you to have the risks worked out as much as possible. The customers agree to take a loss on their investment if things go wrong, and since they pay up front at the beginning of the season, there's nothing they can do about it. If you do operate a CSA in your first year, be prepared to refund people's money if things go wrong.

A CSA operating out of one farm (rather than a co-op) is a challenge because you have to grow so many varieties of food, and they have to be planted in succession so that crops are ready every week of the season. The knowledge and planning required to pull this off comes with time and is incredibly challenging in the first year. It is much better to plant a few high-profit crops, grow several varieties of those crops so you can pinpoint which ones do the best in your area and soil, and raise some animals on the side. Dairy, eggs, chickens, lambs, and goats are reliable moneymakers that can be managed without much more labor. These can be

Example of Weekly CSA Delivery Values

	Week 1	Week 2	Week 3	Week 4
Crops	arugula cilantro cucumbers garlic scapes kale kohlrabi radishes spinach summer squash	beets bok choy broccoli cucumbers dill green onions lettuce mustard greens summer squash turnips	basil cucumbers garlic scape kale kohlrabi lettuce radishes snow peas spinach summer squash	beans beets carrots cucumbers herbs hot peppers lettuce tomatoes summer squash turnips
Value (example)	$26	$26	$27	$28

sold on the farm or at the farmers' market, and are a good, relatively low-stress way to start.

That being said, the CSA is a phenomenal way of creating farm income when there otherwise wouldn't be any. Rather than waiting until July or August to make money, when the crops start coming in, a CSA brings in money in March. This means having cash to purchase inputs like seeds and fertilizer when you need it. In return, the farm promises to provide a share of the harvest to the members each week at a fair market value. This usually averages around four to ten pounds of food per week, since most CSAs charge around $20 to $40 per week. The beginning of the season features more salad greens and cold-weather crops, making the delivery lighter, but toward the end of the year it balances out with heavy crops like squash, potatoes, melons, and fruit. CSAs tend to cater to the more affluent customer who is looking for a more involved experience with the farm, because the upfront cost is usually $300 to $900. These people don't mind picking up their share at the farm every week, or will pay extra for delivery.

It's crucial not to reinvent the wheel with a CSA, especially in the beginning stages. The CSA model is now widely accepted and recognized by customers,

who are willing to bend over backward to help a farm out, including volunteering, handling pickup points, and providing emergency donations. Farms don't have to try to one-up the competition with lower prices or even more services. Keep it simple. Competing on quality and dependability doesn't cost more, as opposed to other ways of competing; it just takes planning and effort.

To calculate your shares, you can create a table like the one below. It's a good idea to start your shares no earlier than June, when you have less risk of crop loss. The value is based on bunches rather than pounds, and arranged according to what is available at the time.

An alternative to the model of pricing shares based on the value of produce (which has become the standard for CSAs in North America) is used at the longest-running CSA, which was formed in 1986. Temple-Wilton Community Farm in New Hampshire calculates the yearly farm expenses and divides that up among the members. Members are then free to collect as much food as they need from the farm. This has evidently contributed greatly to their success, and the waiting list is so long they no longer collect names. Obviously this model forgoes profits in the interest of community.

RECORD KEEPING

"We learn from failure, not from success!"

—Bram Stoker, Dracula

The first step in a microregional food system is collecting information. Many universities have already done work in this realm, which can be accessed through their horticultural programs. Municipal programs may also be a source of information. If this data does not exist, it's time to create it. This information should include:

- plant varieties and species that do well in your area;
- climate information, including frost dates, rainfall, and temperatures;
- current agricultural growing spaces; and
- a fruit tree inventory.

These last two may not be readily available as a quick search on the Internet. If you are creating this database for an entire city, the task could take years. On a small scale, the job requires scouting your neighborhood, or the area you can reasonably travel by walking or biking, to learn what's near you.

Agricultural growing spaces can include:

- backyard and front yard gardens
- greenhouses
- unused wetlands
- parks
- vacant lots
- rooftops
- schoolyards
- community gardens

Google Maps is a brilliant tool that can make this job much easier. This tool is currently known as My Maps, and it can be found by Googling My Maps, or through the menu on the regular Google Maps site. You can create and save a custom map using satellite imagery. Locations can be marked with a pin and with color-coded overlays. Notes can also be added to each marker. Each site record should try to include:

- size and usage possibilities
- soil quality
- water access
- ownership
- possible toxicity

The fruit inventory works similarly in that each tree or bush is marked on the map with variety and possibly notes on yield and harvest dates. In fact, these two groups of information might be more useful on the same map, but that's up to you.

Why is it so important to find or make an inventory of fruit trees in your farm's neighborhood? Observation is the key to success in farming, as is meticulous record-keeping. Your first year of farming is a huge experiment because you are dealing with a microclimate that has no history of farming—except that it might. If you can find fruit trees and other plants growing in your area, you can find out the species that do well and save yourself a lot of experimentation. You might also be able to glean fruit for sale without having to put in much labor to grow it.

Farm Record-Keeping

Record-keeping begins with planning, but it doesn't end there. You will need to keep detailed records of:

- all your plant varieties, including what seed brand (or source) you got your seeds from, and how many you used;
- any problems per plant;
- when you started harvesting and how much you got;
- notes for yourself for next year;
- your expenses and sales.

This will make it possible for you to improve each year and produce more based on your past failures and successes. Only one in five small farms is keeping

▼ This suburban land behind an RV park was borrowed in exchange for food, and it had no soil. Soil was brought in and held in place by wood boxes.

records of their farm production. Many farmers don't want to see their farm as a business and would prefer to keep it as a social enterprise. That's great, but even if you are operating without profit you should record your growing information. There is no way you will remember which type of tomato did well and which didn't, or whether you remembered to fertilize, or when you started your mushrooms. Many farmers now use a simple spreadsheet. If you have a smartphone, most spreadsheet software can be saved to iCloud and will be accessible on your computer at home or anywhere you go, so there's really no excuse for not writing down how much garlic you grew.

ACQUIRING LAND

"As soon as the land of any country has all become private property, the landlords, like all other men, love to reap where they never

sowed, and demand a rent even for its natural produce."
—*Adam Smith,* Wealth of Nations

Acquiring Land

There are a lot of ways of getting land, but not all areas are ideal. For commercial agriculture in the city, your crop will sometimes have to change in order to work with in the amount of space you have. For example, if you live in an apartment, you will not be growing cabbage as a farmer. You could, however, grow herbs, microgreens, mushrooms, and other delights. For urban farmers, the options are:
- leasing land from a land trust or the city;
- leasing land from wealthy (or not so wealthy) nonfarming landowners who get a tax break (in some states and provinces agriculture receives a tax break, but this only applies if a percentage of the land is *actively* growing something);

- renting a portion of a working farm in a suburb and sharing equipment;
- farming land owned by a school, restaurant, retreat center, or other institution;
- borrowing an urban backyard from a willing person who might be willing to barter for vegetables;
- farming a rooftop with corporate partners;
- renting urban land from the city, such as an unused park space or empty lot;
- farming on the site of an old bedding plant nursery or other compatible space;
- growing indoors under lights.

Buying versus Renting

In the city, land prices are much higher per acre. In the area where I live (Victoria, British Columbia), prices run around $300,000 per acre, and this includes rural areas. Typically, people run to the country and buy cheap land. The benefit of being in the city, however, is the market. Your customer is literally right next door, which is a much more profitable situation. At the same time, you need to have space. If you don't have the money to purchase a property, you will have to rent or borrow it, but you shouldn't let that stop you. Acreage is nice to have, but not even totally necessary. However, if you can buy property in a feasible way, do it. You'll protect your investment and you'll have much more control over what you can do. If you can't, there are many other ways of getting land.

Borrowing Land

More often than not, urban farmers use someone's underused hobby farm acreage in the suburbs. This land is either leased or borrowed. However, leasing with a contract may not necessarily have any additional security over a verbal agreement, since the farmer is only leasing a section of the property. The property owner can still probably kick you off at any time. At the very least you might be able to get a deposit back or retain your property, such as tools or a greenhouse. Without a paper agreement, the property owner can take everything you have built up there. Verbal agreements only work well if the person you're dealing with is trustworthy, but even then, having something in writing is a good idea.

Loss of farmland is the biggest complaint of urban farms, and this is why their farming techniques are slightly different. For borrowed or leased land, raised beds and permanent structures take too much time and money to invest in, unless it is absolutely certain that the property can be farmed for the next five years. In five years it will have paid for itself. Otherwise, the property must have soil that is already good enough for farming, and a rototiller is used to turn the soil. The farmer must be able to just add fertilizer, turn the soil, and produce a crop right away without much investment in boxes, fences, and other structures.

A pitfall of borrowed land is toxins. Urban farms in these areas are not usually certified organic because they do not have adequate buffers. Gardens growing on the edges of roads and highways have a higher lead content because of deposits in the soil, and this is why certified organic crops must have a buffer zone, either of a certain amount of space or a tree belt to absorb the exhaust fumes. To combat this, urban farmers must add even more compost to the soil. The more organic material, the less concentrated the lead and other toxins will be. Mulch can also help.

Benefits of Close Proximity

One major benefit of urban farming is the close proximity to customers. Some even have farm stands set up right on the sidewalk like lemonade stands. Most cities have a plethora of farmers' markets, so selling the produce isn't the problem; it's growing enough to meet demand, or growing the right things and finding the right market. This possibly means using several backyards within a small area. However, too much traveling makes crops suffer because they cannot be properly cared for. The multiplot model requires drip irrigation on a timer, which is more expensive than spray irrigation but saves the trouble of running around to all your plots four times a day for watering.

The urban farmer needs one permanent spot to process crops. This means harvesting everything and then bringing it back to a place that has tables, a water supply, and refrigerators or a walk-in cooler. An urban farm cooperative can help put this together; otherwise, the farmer will have to have a garage for processing. This processing area needs to be close to the farmer's plots as well.

All of these are fairly standard farm issues, but the real challenge is zoning. City farms are facing conflicts, and unless the city has specifically allowed urban farming, it is probably illegal. The urban farmer is making the conscious choice to break city bylaws in the hope that the neighbors won't complain. There are two choices here—a tall fence can hide the backyard and make it impossible for anyone to even know what's going on, or the farm can embrace

the publicity. If the second option is chosen, be prepared to meet all the neighbors, offer them free vegetables, and become an ambassador for farming.

Some factors to consider when finding a location:

In the city it's not very likely that you will have much choice in what properties are available to you. However, you can decide where your home base is and whether or not a property is going to be worth farming on. You should ask yourself these questions:

- Are there employment opportunities? Be prepared to keep steady employment while you build up your farm business. If work is close to your farm, it makes it easier.
- Is there a good sense of community? Do they accept new people?
- Is there culture and education? Communities with culture and learning opportunities have tended to embrace urban farms, and the two seem to go hand-in-hand. This provides connections to possible customers, as well as emotional support for you.
- Are there churches and other community places located in the area? These are one of the best free resources for farmers to use as distribution points or to get the word out.
- Is there a farmers' market or grocery nearby?

As I mentioned, community is a major component of urban agriculture. Some urban farms have been successfully put together in some inner-city areas and these farms have been able to bring the community together in amazing ways. However, as a farmer with a business, it is

▲ This may not look like prime farmland, but farmers in Detroit think otherwise. Land like this is being reclaimed for farming.

a lot easier to begin in an area that already has a strong sense of community, lots of culture, and a decent farmers' market. You will have to make a judgment on how much energy (emotional and physical) you want to put into building your community up.

A Conversation with Your Landlord

As a renter, you have many options. Most landlords are fine with a small garden bed. It's a tough sell to convince your landlord that completely tearing up the entire lawn is a good idea. Your landlord will be concerned with:

- liability (Could there be some fallout on them because of this activity? Does their property liability insurance cover this?)
- the value of the property (Could it decrease?)
- water (Who pays the water bill? If it's you, no big deal, but if it's your landlord they will see a huge increase.)
- neighbors (How will they react?)

Your landlord is concerned with risk and investment, which should be your concern as well. This means now is the best time to write a preliminary business plan based on the property you are looking to farm on. Perhaps you had visions of an orchard or blueberries, but the land isn't quite right for it. The property will shape what you are doing to a large extent.

If you are using your own rented house, finding land isn't the problem, it's getting permission to use the land. But if you are looking for land, you still need to identify whether it's a suitable place for farming. Just because it's empty land doesn't make it the best spot.

Here is a site checklist:

- **Ownership:** Write down the address, including the cross streets. The owner will be listed at the county or provincial tax assessor's office. Sometimes this is available online.
- **Water access:** If there is no municipal water service or a well, installation will cost you thousands of dollars. As mentioned before, this is not worth it

unless you can guarantee access to the property for at least five years.

- **Transportation:** The property needs to be easily accessible by truck. Either you need to bring in supplies, or you will need to get a truckload of food out.
- **Light:** The sun needs to be shining on the ground for at least 60–70 percent of the day, or you will need permission to cut down trees. If a building is shading it, it's a no-go.
- **Wind:** It should be sheltered from heavy winds by trees or buildings. This can sometimes be remedied by building some windbreaks.
- **Drainage:** If the soil is flooded or marshy, there is no easy fix other than some major draining construction.
- **Soil:** Is there soil? What is the quality? Could it be contaminated? Cities have a ton of concrete and contaminated soil. This doesn't eliminate the property, but it does play a big part in your planning process.
- **Time:** How long will you have the land?

Some of these factors will be addressed in this chapter and have solutions. Some of them cost too much, so you'll have to decide whether it's worth it.

Making a Farm Plan

Here is a farm plan outline:

- Executive summary
  - a brief summary of the business
  - your goals and timelines
- Layout diagram of the farm property
- Benefits
  - to the community
  - to the property owner
- Management: your own experience and personal references

- Construction:
  - changes to existing structures
  - development of new structures
- Maintenance: property maintenance measures
  - safety
  - cleanliness
  - aesthetics
- Termination of Use: cleanup and restoration of the property
- Operations:
  - who has access to the property
  - who will work the land
  - who will pay the water bill
- Licensing & Insurance:
  - business licenses
  - agricultural inspections
  - liability insurance

You will need to write a cover letter, which will be a brief introduction to who you are; a proposal to lease the land for free (or for a low price); your willingness to pay for any expenses that you write about in your farm plan; and a big thank-you for their time. Most landlords will not be receptive to animals, although some municipalities have been accommodating if you are using land owned by the city.

This preliminary plan is a good starting point for your own business plan, and it gives enough information to bring in a potential landlord. It does not include your operating costs and income projections, which we will talk about later.

Developing a Written Agreement

A written agreement to use land for farming *needs* to be in writing and signed by both you and the owner. This should be done no matter what has been discussed. This must include the following:

- the name and address of the owner and his or her spouse

- tax information if the owner is a business, as they would have to pay tax on the rent
- the name and address of the renter(s)
- address of the land
- a map of the land, with the boundaries clearly detailed
- the length of the agreement in months; the minimum should be twelve months
- an option to renew
- how much notice is needed to terminate the agreement and under what circumstances
- an agreement that the renter cannot sublease
- explanations of what happens if the land is sold or the landlord dies
- outline which areas of the property the renter will have the right to access
- hours of the day the renter is allowed to access the rented areas
- what kinds of exceptional circumstances might cause access to be changed
- who the renter is allowed to bring on the property and why
- if members of the public are allowed in the area and with what notice
- the specific business activities of the renter
- activities that are prohibited
- how manure is managed
- if farm signage is allowed
- what production practices must be followed in order to maintain soil quality
- any certifications or regulatory constraints such as organic certification, agricultural land status, or tax benefits
- any exclusive use of any buildings or structures and for what purpose
- any shared use of any buildings or structures and for what purpose
- bathroom facilities available to the renter
- who is responsible for building maintenance
- if fencing is required by the renter, who is responsible for maintenance
- if the renter plans to build permanent structures, whether there will be compensation
- use of electricity and monitoring of usage
- use of equipment owned by the landlord and compensation
- water sources and use, backup plan in case of drought, and who is responsible for maintenance

If a house or residence is offered as well, this use should be put under a separate lease agreement.

Compensation to the Owner

Many urban farmers have negotiated the free use of land in exchange for food. For example, a backyard can be borrowed from a homeowner in exchange for a box of food every week. A large plot of land, for example, an acre on a hobby farm in the suburbs, might be a different story. If a long-term commitment is being offered by the landowner, some compensation may be fair. However, what is considered "fair market value" is not necessarily the best guide for a piece of acreage. You will have to find a willing person who is happy to let the land go to use for farming without much compensation. You can, however, compensate the farmer for water, power, and other amenities, with a small amount thrown in for using the land.

▲ This garden is located at a hotel and supplies the restaurant with high-quality herbs and vegetables.

BUSINESS PLANNING

"Give me six hours to chop down a tree and I will spend the first four sharpening the axe."

—*Abraham Lincoln*

Scale

Typically, the definition of a farm is acreage that is a certain size and makes a certain amount of money. However, that definition doesn't apply to urban farming. There are many different scales of agriculture, and they don't necessarily have to do with size, but rather with intensity:

- *Microgarden:* 1,000 square feet or less, usually supplies one family with vegetables.
- *Yard:* 10,000 square feet or less, this amount of space can feed a whole family most of the year plus garner surplus for sale.
- *Lot:* 10,000 square feet or less, as above, except that it is not attached to a house.
- *Intensive:* 10,000 square feet to 1 acre, either on a lot or small acreage.
- *Small farm or food garden for school/ institution:* a food garden attached to an institution operated for education and nutrition rather than profit.
- *High-yield farm:* minimum of 25,000 square feet in full intensive production.

In an urban setting, the maximum amount of land that is usually farmable by one farm manager is two and a half acres, but more land is not always more profitable. Curtis Stone of Green City Acres had two and a half acres and eight staff members, but found through careful record-keeping that ten crops were making 80 percent of the profit on one acre. Does this mean only grow the most profitable things? Not necessarily. Farming sales often use "loss leaders," or low-priced, low-profit crops to boost sales of more profitable crops. Fifteen vegetables of multiple varieties on one acre can produce a living wage for a farmer, with less overhead and less time spent working. The goal (as Curtis has also demonstrated) is a forty-hour workweek with a payout of forty to fifty dollars an hour.

This is done through high-rotation, intensive succession planting. These crops are planted in beds of a standard size: on a suburban lot, usually thirty inches wide by twenty-five feet long. When initially creating the bed you would need a sod cutter to remove the sod, add amendments (fertilizers and other similar stuff), fork it up to break up chunks and loosen up roots, then rake it out to remove debris and invasive grass. Each standard bed would be seven hundred fifty square feet.

You would have two kinds of beds:

High rotation: Three to six rotations per season, located close to home with a crop value of $800.

Bi-rotation: Two rotations per season, $400 crop value.

The next chapter has more information about preparing these beds using either a rototiller or no-till method. During your planning, however, you would want to decide what you want to grow first.

In order to create a steady stream of income, you would plant two kinds of crops:

Quick crops: These crops are ready in fourteen to sixteen days. You would plant them in succession weekly because you harvest them once rather than picking them over time like tomatoes. A sample schedule will be discussed later. These include turnips, microgreens, radishes, salad greens, beets, green onions, and carrots.

Steady crops: These are ready in sixty-four to eighty-five days, and you would plant them once (maybe twice) because they produce a steady crop throughout the season. These include patty pan squash, indeterminate tomatoes, peppers, eggplants, kale, beets, and carrots.

Profitability Chart for 750 Sq. Ft. Beds

Type	Amount	Time	Growing	Income
Radish	80 bunches per bed	30 days	7 rows	$200
Patty pan squash	60 lb. per bed	130 days	Greenhouse starts, interplanting, succession	$240/$4 lb.
Field microgreens	10 lb. per bed	15 days	6 ft. bed	$150
Leaf lettuce	15 lb. per cut	70 days	3 cuts per bed, 45 lb.	$360

Calculating Your Gross Sales

If you have twenty-four high rotation beds at $800 and twelve bi-rotation beds at $400, you would make a $24,000 profit.

You also need to calculate your costs. If you get a new plot of land and put in a bunch of time and money, it needs to pay for itself. Let's suppose you have a space that is around two thousand square feet. This allows you to put in two beds, with enough space for walkways. According to Curtis Stone, you would spend $978 to put in two beds (see the table below).

This means that these two beds will pay for themselves, plus make a small profit the first season. In the second season, inputs would be your only expenses.

More Models of Farming

Besides the scale of the farm, there are a few more types of urban farms that need to be addressed:

Indoors: An apartment could be an indoor farm. A warehouse can also be an indoor farm. Growing under lights can be a viable option for many different crops, as long as the energy cost is factored into the cost of production. If you are an indoor farmer, the discussion of land and planning may not apply to you, so just skip ahead

to the indoor growing sections. Crops that work well inside are mushrooms, greens, microgreens, and herbs because they are quick-growing or don't need quite as much light as other plants.

Windows/balconies: Although this doesn't allow for much farming space, certain crops work well on balconies and in window boxes and can give you a tidy cash crop. Tomatoes, dwarf fruit, hardy berry bushes, and mushrooms can give you high cash returns from a small space.

Community gardens: Most community gardens do not allow you to farm commercially (that is, to sell what you grow). However, if you want to use a barter model, a great idea would be to grow a ton of food and trade with the other members of the community.

Farm to table: Farm to table refers to farms that specifically sell to restaurants. Some of these are located at the restaurant itself. Others specialize in fresh specialty crops and deliver directly to discriminating chefs. This is usually very profitable.

One person alone can't maintain a productive farm that is growing a serious cash crop. An entire family has to be involved, or a group of seasonal helpers. There are several factors that can help you

Item	Cost	Total
Labor: sod removal, rototilling, forking, fertilizing, raking	$15 hr.	$450
Sod cutter rental	$60 per day	$60
Fuel expense	$20 per day	$20
Irrigation, timers, impact heads, lumber, poly lines	$320 one-time purchase	$320
Soil inputs: 2 yards compost, 2 cans fertilizer per bed	$80 compost $48 fertilizer	$128
	TOTAL	$978

choose a cash crop. As a product it should take up less space, such as honey or berries. It should also be easy to process without much equipment, or be *value added*. Value-added products are jam or butter—raw materials like berries and milk that are processed into something people want, which adds value. It's a good idea to have a second cash crop that is something nonperishable, like firewood or nuts, which can be sold throughout the year. Here are some cash crop ideas:

Aquatic nursery: Fish, bee, and duck forage; friendly insect plants; and ornamentals

Berries: Berry sales, U-pick service, plant nursery selling berry starts

Rare plants: Useful permaculture plants; bee, bird, and beneficial insect forage

Seeds: Rare or unusual heirloom seeds

Animals: Geese, silkworms, earthworms, bantams, milk or mowing goats, specialty sheep, and quail

Hedges and trees: Local tree species, regenerating forests, windbreaks, animal forage, bamboo, or food crops. (These products can be grown in containers, which make them possible in an urban setting.)

Organic food: Typical fruits, vegetables, nuts, milk, eggs, wool, meat, flowers

Value-added food: Smoked meat, dried fruit, jam, feathers, dried flowers or wreaths, pickles

Craft supplies: Willow, bamboo, natural dyes, wool

Natural pest control: Nursery plants like marigold or yarrow

Herbs: Dried herbs for culinary or medicinal purposes

Commercial Greenhouses

It is much cheaper to build a greenhouse than to buy one, but take into consideration the work and potential risks if you do. It is common for businesses to start out by building a *high tunnel*, which is simply a long framework made of PVC piping that holds up a light plastic cover. This plastic cover must be removed in the winter, so this is not for year-round growth. This type of greenhouse is cheap, but it is also very susceptible to wind damage.

It is much easier to buy a real commercial greenhouse, which will have built-in features like ventilation and be sturdy enough to use all year round, than to build one. One feature it needs to have in a cool climate is a separate entry room. Every time the door opens heat is lost, so an entryway with two doors is crucial. The first crops you grow should be designed to pay for the greenhouse, which means growing high-value products like tomatoes, peppers, and fruit.

Of course, while running a commercial greenhouse business is most often done with a single high-value crop, in a permaculture system you could still follow the principles of community growing and polyculture. Pick at least three crops that can work together, and consider using aquaculture tanks or birds as well.

Planning for a CSA

Many studies have been done on CSAs, and what you can expect is somewhat predictable. A traditionally farmed row-crop farm using mechanization can expect to produce thirty shares per acre. That doesn't mean that the acre is going to just feed thirty families; it just means that as a general rule, planning for losses, no more than thirty units should be sold. An intensively planted farm can offer up to fifty shares per acre when well-established, but is unlikely to be able to farm more than two acres without a lot more labor.

Intensive farming requires at least five people per half acre to take care of the weeding and harvesting, and the more land, the more people it takes. At the two-acre level, the farm should have anywhere from eight to twenty people involved at various times of the year. A mechanized farm can use five people for five acres.

The farm must plant at least thirty to forty varieties, and each share is built around a large salad. In some areas a mixed baby green bag is expected (mesclun mix), and in others a large head of lettuce is the norm. Early in the season things like broccoli, kale, cabbage, and turnips fill in the weight, and later in the season the tomatoes, zucchini, cucumbers, and potatoes fill it in. Baby greens are a challenge because a new crop must be ready every week. This means planting every week for harvesting thirty to thirty-seven days later.

Once the maximum farm customers are obtained according to the space you have, the next step is *not* expansion to more land or trying to get more customers. It is more profitable to maximize production and create more value. Value-added products are ones that are processed from farm crops, such as jam, pickles, and tea. Not only can you make many of these items from *seconds*—that is, crops that are too ugly to sell—you can charge more than you could for the raw item. These can be sold as add-ons to your existing customers, along with eggs and meat, and can be sold at your farm stand. Meanwhile, work hard to build up your soil and perfect your growing techniques for greater and more reliable yields.

Distribution can be done by home delivery, delivery to drop-off points, and/or pick-up at the farm. Each has its merits and drawbacks. Home delivery puts distribution completely in the hands of the farm, but takes more labor. For the farmer all the customers should be within a thirty-mile radius, otherwise it's not worth the fuel and time. It takes less time to do this than to allow farm pick-up, but you have to have a very good delivery database system on the computer that keeps track of the customers, the routes, and whether they will be around or not. Deliveries also have to be made in the evening when people are home. Delivery to drop-off points is a happy compromise but usually relies on an outside source to provide the location, such as a parking lot or even a member's garage. Many farms have a relationship with a member that allows them to drop off the food at a spot and leave, and allows shareholders to pick up when they can during a three-hour period. Pick-up at the farm works if the farm is close to the shareholders and has ample parking. The shares can be left out and people can serve themselves without too much trouble.

Example: CSA Availability Chart

Variety	May	June	July	Aug.	Sept.	Oct.
Greens						
Arugula (baby)						
Endive (baby)						
Escarole (baby)						
Kale						
Lettuce (baby)						
Mizuna (baby)						
Nasturtium						
Spinach (baby)						
Herbs						
Basil						
Cilantro						
Dill						
Parsley						
Sage						
Legumes						
Bush Beans						
Snow Peas						
Roots						
Beets						
Kohlrabi						
New Potatoes						
Radish						
Carrots						

Example: Crop Planting Schedule

April 29 – May 5

Plants to Be Planted

Plant	Location	Yield Needed per Week from Group	Seeds Required per 25 Sq. Ft.	Estimated Yield per 25 Sq. Ft.	Alloted Sq. Ft. Required	Days Remaining till Mature	Expected Harvest Date
Radish (small)	Y3	150 bunches	895	25 bunches	150	24	May 24
Carrots	E10	75 lb.	750	25 lb.	75	60	June 25
Onions (bunching)	G9	75 lb.	1250	25 lb.	75	65	June 27
Nasturtium	K7	10 lb.	25	3.5 lb.	75	60	June 25
Kohlrabi	L7	40 lb.	375	12 lb.	100	42	June 9
Potatoes	M8	100 lb.	2 per can	50 lb. per can	2 cans	60	June 25
Escarole (baby)	S5	10 lb.	150	3.5 lb.	125	55	June 11

Plants in Growth

Plant	Location	Yield Needed per Week	Seeds Required per 25 Sq. Ft.	Estimated Yield per 25 Sq. Ft.	Alloted Sq. Ft. Required	Days Remaining till Mature	Expected Harvest Date
Leeks	A1	350 stalks	150	50 stalks	175	17	May 15
Leeks	A2	350 stalks	150	50 stalks	175	24	May 21
Leeks	A4	350 stalks	150	50 stalks	175	39	June 4

Plants to Be Harvested

Plant	Location	Yield Needed per Week From Group	Seeds Required per 25 Sq. Ft.	Estimated Yield per 25 Sq. Ft.	Alloted Sq. Ft. Required	Days Remaining till Mature	Expected Harvest Date
Bush Beans	B1	40 lb.	200	20 lb.	100	0	April 20
Peas	C1	40 lb.	200	20 lb.	100	0	April 30

Developing a Final Business Plan

You've found your land, have a good idea of the type of farm you want to have, and a general idea of the crops you will be growing. The next step is to write it all down in a business plan that will help you maintain your goals and keep you on the right track. The simplest farm business plan is outlined like this:

- **Description:** Who are you, what does your farm grow, and how do you grow it?
- **Land:** Who owns the land or building, and how do you protect the owner and yourself? The section earlier in this chapter on acquiring land includes a lot about how to work out the legal use of your space, and it will be helpful when writing this part of your business plan.

- **Income:** How much do you need to grow in order to make as much money as you want to make? These are your objectives and also gives you a healthy dose of reality. Does what you want to make jibe with what you have?
- **Niche:** Just because one of these crops seems more profitable doesn't mean that focusing on one crop is a smart idea. Diversifying and bringing in customers with a cheap bunch of radishes is how most urban farmers become successful. How will you market and sell your crops?
- **Costs:** The best way to be profitable is by reducing costs, and the biggest

one is labor. Inputs like seeds and fertilizer are a close second. How can you manage your costs and still maintain quality?

To finalize your business plan, you must make some predictions on your expected crop yields. The following small farmer's table includes average yields as a guide to predicting possible profits. It includes some crop yields and retail prices per square foot, with retail price range per pound. This range is taken from a variety of sources. The best place to get accurate information, however, is to go to your farmers' market and find out what farmers near you are charging.

Crop	Average Yield per Sq. Ft.	Price Range per Lb.
Arugula	2 lb.	$9.50–$10.00
Basil	½ lb.	$5.75–$10.00
Beans	¾ lb.	$1.00–$1.50
Beets	1 lb.	$2.50–$3.99
Broccoli heads	½ lb.	$1.00–$1.75
Brussels sprouts	½ lb.	$1.00–$1.90
Cabbage	1¾ lb.	$0.75–$0.80
Carrots	3 lb.	$1.00–$3.00
Cauliflower	¾ lb.	$0.50–$1.25
Chard	4 lb.	$3.00–$6.00
Cilantro	¾ lb.	$4.00–$8.00
Cucumbers	3 lb.	$2.00–$3.00
Dill	¼ lb.	$3.75–$5.00
Eggplant	1 lb.	$3.00–$3.50
Garlic	¼ lb.	$8.00–$10.00
Greens, baby	2 lb.	$6.50–$7.00
Kale	2 lb.	$4.00–$8.00
Lettuce, head	1 lb.	$0.75–$1.00
Onions	3 lb.	$1.00–$2.00
Parsley	½ lb.	$3.00–$6.00

Crop	Average Yield per Sq. Ft.	Price Range per Lb.
Peas, sweet	1 lb.	$1.00–$2.50
Peppers	¾ sweet, ¼ hot	$3.00–$4.00
Potatoes	1¾ lb.	$2.00–$3.75
Sweet potatoes	1½ lb.	$3.25–$3.50
Spinach	1 lb.	$2.75–$8.00
Squash, summer	2 lb.	$0.50–$2.50
Squash, winter	1 lb.	$0.50–$2.00
Tomatoes	2 lb.	$2.50–$5.00
Turnips	2 lb.	$2.00–$3.00

FUNDING

"Most business meetings involve one party elaborately suppressing a wish to shout at the other: 'just give us the money'."

—*Alain de Botton*

The current agricultural system is dependent on petroleum at every stage of the food process. Fertilizers and pesticides are petroleum-based. The tractors run on petroleum, and the food that is produced is transported using petroleum. All of this creates carbon emissions, and agriculture is responsible for a lot of it. The EPA estimates that 9 percent of all greenhouse gas emissions comes from production, and this doesn't even include the transportation. Changing the way food is produced and distributed can have a major effect on humanity's future.

However, agribusiness has effectively blocked a lot of change. This means that funding for your small farm is not necessarily going to come from the government or a bank as it would for any other business.

In doing research for this book, it was interesting to hear the different opinions on how to fund young and beginning farmers. When we describe "young" farmers, we now refer to something quite broad—currently, farmers make up only about 2 percent of the population. They are mostly in their sixties and are trying to retire, and so any young, beginning farmer is really anyone under sixty.

The government is totally supportive of farmers who are interested in agribusiness, and this is the case in Canada as well as the United States. If you should decide to grow soybeans or another commodity crop on large acreage, you are likely to have many funding doors open to you. However, the typical beginning farmer is a woman looking to start a small, organic farm or market garden. Often these are not "young" people, but rather mothers, graduate students, and immigrants who have a lot of life experience and education. The funding options are far fewer and take on the form of incubator farms, microloans, and land leasing. Incubator farms are properties owned by an organization that lends the farmer land, training, tools, and other inputs to start a farm at no cost.

Organizations that support incubator farms have described a situation wherein they have supported new farmers only to find that the new farmer is unable to become independent of the incubator farm. They have found that without the cooperative funding model, new farmers today really struggle to create a reliable income. Young farmer coalition groups and organizations have described a similar situation and would like to see a better funding model from the government that treats all farms as a necessary system like transportation or health care.

Regardless of what the future holds, it is important to realize that if you are going to make a living farming, you will be dependent on a community of people to make that happen, whether it be a cooperative farm, an incubator farm, a land lease with a benevolent owner, or through networking within your community. It is hard to find examples of sustainable, small-scale urban farming that provide someone with a full-time living. They do exist, and those who are making a living are often selling the information on how to do so. It's important to do your homework and have a solid business plan that can outline your farm income in a realistic way and help you figure out how much income you can comfortably produce. At the beginning, you will have to have other income and a plan to scale up your production so that you can make your living as your business grows.

2 | Designing Your Farm

PERMACULTURE

"You can solve all the world's problems in a garden."

—Geoff Lawton

Planning with Efficiency and Permaculture in Mind

Although permaculture principles are a super helpful ideology before you start farming, designing the actual growing space for efficient production on an urban farm using permaculture is a lot more difficult. There's often not a lot of choice in where things are in a city, and you can't do the long-term development that permaculture requires. However, permaculture is a useful tool in the initial planning stages using a system of *elements*.

These are the elements of urban farms:

- intensive production beds or containers
- light source
- rainwater collection
- waste disposal
- food processing
- small orchard
- greenhouse
- duck, chicken, and goose area
- goat shelter and grazing
- beehives

Kinds of production:

There are also different categories of production that determine the location and type of bed or container that you use to grow it:

Human food: The options are niche market crops or straight-up vegetable production.

Cooking and medicinal herbs: These are often hardy perennials that deter pests.

Animal food garden, or forage: Rather than letting a field lie fallow, a space-saving and cheap solution can be to plant a cover crop such as alfalfa and allow the animals to forage. Alternatively, you could plant a garden that not only feeds chickens, but attracts insects away from your vegetable garden so that the chickens eat your pests.

Soil fertilizing, green manure: A ground cover like alfalfa not only saves topsoil, it adds nutrients to the soil. Microgreens are a profitable way of creating green manure.

Insect control: There are a variety of plants that deter pests. Rimming your gardens with them is a vital method of increasing the yield of your vegetable production. Many of these are also edible herbs.

Locations

Animals: Birds should be located near a vegetable garden because they can help keep away pests by eating them. However, they will munch on your vegetables, too. So if you use this method, plant a lot of extra food. Goats should be located away from gardens you eat from, but near grain sources that they can forage on. Fish go in the pond, and bees go near gardens that flower, such as the orchard.

You: If you are using borrowed suburban yards, your most intensive crops (that is, the ones that require the most work) should be located nearest your home.

Water: Rainwater will most likely be harvested off shelters and living areas, and so will be near where it is needed. Water can also be used for passive solar reflection in greenhouses.

Orchard: Orchards are rare in the city, but if you do have one, it should be used as a windbreak and to prevent soil erosion.

Waste: Manure from animals should be covered, and in its raw form should not be allowed to leech directly into the ground next to food crops. It needs to be composted first.

Measuring Success

For the land to be considered sustainable, it needs to produce at least as much as or more than it consumes. In a permaculture system, success is measured by the energy that is available in the system and the intangible benefits that are reaped. However, urban farming must, by necessity, measure success through pounds of food the land can put out. There are some other measurements that urban farmers can enjoy:

1. Production increases when irrigation is steady and the soil is healthy. Roots can then penetrate deeper and get essential nutrients.
2. Using gravity-fed water saves energy and can be used to recycle water. Electricity usage is a measurable energy (and money) expense, and the less you use to pump water, the better.
3. Providing tree windbreaks and forests for animal forage over 20–30 percent of the land will increase production simply by providing shelter and microclimates for plants and animals. They will also provide food for themselves, and homes for predators that eat pests.
4. The farmer, despite *not* having a tractor, has to perform less physical labor because of the interconnectedness of the system.
5. The land has recreational value for people to enjoy.
6. Future generations can enjoy the land.

One way in which I would suggest deviating from true permaculture is in cultivating ecological succession. Typically, a permaculture system would recommend creating a self-sustaining growing pattern that would mimic the pattern of the forest, or creating an *ecological succession.* The end goal would often be a food forest, which takes ten years or more to develop. Since we don't have the luxury of that much time, we will still use the commonsense design approach permaculture provides, but without the focus on some of those long-term projects.

Permaculture Words to Know . . .

Here are some important words that are unique to permaculture design. Other systems use some similar words, but permaculture uses them differently.

Polyculture: Monoculture is the type of picturesque farming that one typically sees in North America, with long rows of a single crop grown over a large flat area. The "amber waves of grain" in America is monoculture, and the complete opposite of polyculture. Polyculture is when varieties of plant and animal species are mixed together for mutual benefit. Orchards can be clumped closely together rather than spread out in neat rows, and they may have herbs and ducks under them. A climbing plant can be grown with a tall plant, such as corn.

Aquaculture: Water systems have the potential to produce much more protein per square foot than an equivalent area of land. A successful aquaculture system is patterned after productive land-water edges such as swamps and coral reefs. Although a permaculture ideal would be a natural pond, urban farming necessitates an aboveground tank with plants growing off the dredged fertilizer and looks very similar to hydroponic systems: in fact, these systems are called *aquaponics.*

Elements: Any feature on a piece of land is an *element*, whether you intentionally placed it there or not. This could be a clump of trees or herbs, a pond, or a pile of rocks. These elements are part of an overall design, each one thoughtfully used to make the land more productive. Each element is also connected to everything else in as many mutual beneficial relationships as possible. This wide variety of connections, or *diversity*, increases efficiency and places value on elements that formerly might have been seen as annoyances. Every "problem" can be turned into an advantage. For example, weeds are simply Stage 1 plants (or pioneers) and can be turned into mulch (if they haven't gone to seed yet).

Design Problems

Almost no urban agricultural land is ideally suited to farming. You will be presented with many design problems, but nothing can totally deter you from using a piece of property for farming except for extremely high levels of toxicity. While traditional farming would say that gravel or hills would exclude the land from being farmed, urban farmers think differently. Some of the most common urban land issues include the following:

Gravel or covered in asphalt: Many successful farms have been placed on former parking lots or former gas station lots. This has been solved by using growing boxes filled with soil from somewhere else.

Hilly: Hillsides all over the world are farmed through terracing.

Cold climate: Winter growing in North America has become a standard practice even in temperate zones thanks to plastic high tunnels.

No water: This is the one issue that is hard to fix. Traditional farms would drill a well or put in irrigation lines, which you can't do in a city. Rainwater collection and gravity-fed irrigation from tanks can solve this.

Toxicity: Some soil has lead in it. The danger from this often has more to do with tilling and breathing dust than lead poisoning from vegetables, but lots of compost is the solution.

Elements as Goals:

Before you move on to the next phase of design, you should first think about your goals. What do you want the land to do for you? What do you need to live? Write these goals down, and turn them into elements:

A way of growing food in winter = a greenhouse

A solution to drought and irrigation = water storage

A way to stay organized = a tool shed

A place for animals to live = a barn

A place to grow food = a garden

A place for food and yard waste = a compost pile

A source of protein and manure = a chicken coop

A way to stop the winds = a windbreak

Inventory of Elements

Now that you have a list of goals and elements that you *want* to have, you need to take an inventory of the elements you already *do* have. Any characteristic of your land, whether you see it as positive or negative, must be written down. This includes large rocks, hills, marshy places, existing structures, trees, etc. Of course, to create a map like this you will have to carefully inspect every corner of the land and use every sense you have to observe the finer details. This may mean taking notes for an extended period of time. First, walk the perimeter of your property and measure how many square feet you have available. The easiest way to do this is to measure the length of one of your normal steps with a tape measure, walk the length and width of the property, and multiply those paces by how many feet one of your steps is. Then take note of these things:

- temperature changes from one area to the next
- how much effort it takes to go from one place to another (for example, from your home to the plot)
- where prickly plants tend to grow
- where insects seem to group or swarm together
- where different species of trees are growing, and the conditions
- how water moves across the land when it rains or snow melts
- which trees have been shaped by the wind, and the direction the wind comes from
- where the sun is warmest and where shadows move
- signs of animals moving, eating, and sleeping
- plants that produce fruit before other plants
- poisonous or unpleasant plants
- erosion or ditches caused by erosion
- damp or boggy ground, which may yield peat or clay
- dead wood or logs that can be harvested

- slopes and hills, their height and which sides are sunnier
- cliffs, rocky places, and rough areas

Categorize all the resources you have mapped and list them in three groups: life, energy, and social. Life resources are the plants, animals, and insects growing there. Energy is the potential wind, wood, water, or gas energy you can use. Social resources are the teaching, recreation, and gathering possibilities for people. There are also resources off the land, such as restaurants and markets for selling products, sawdust from sawmills, schools, and a population of people who might be a potential market for your farm produce.

Take time to analyze each element. Think about what it needs to live and what it needs to be useful. What are its characteristics, its behavior? What does it produce and what beneficial functions does it provide?

Categorize these by characteristics, input, and output. Input is what the element needs to function; anything that allows it to be useful or keeps it alive. Output is what it can provide to other elements. Characteristics are the attributes that make that element what it is. Make sure you number the element for reference later. It is easiest to write each element on a three-by-five card:

1. Rock Pile

Characteristics: Dark slate

Input: Sunlight (to radiate heat), human labor to move the rocks around

Output: Insulating and passive solar properties, structural support, roofing, flooring, wall for passive solar heat, windbreak, shelter

2. Herb Garden

Characteristics: Variety of perennial herbs (comes back every year)

Input: Sun, water, mulch

Output: Medicinal herbs, cooking herbs, bee forage

3. Chickens

Characteristics: Barred rock chicken, dual-purpose breed for both eggs and meat

Input: Food, water, grit, shelter, other chickens

Output: Eggs, meat, feathers, manure, methane, foraging

4. Deer

Input: Food, water, forest, other deer, marked territory/path

Characteristics: White-tailed deer

Output: Meat, manure, methane, foraging

Once you have taken an inventory of all the elements you have and that you plan to build, you will need to match the outputs with the inputs of other elements. For example, if bees need to eat and the herbs can provide food for them, they should be planted in proximity to the bees. The chickens need food every day, so the coop should not be too far from the house. At the same time, they produce manure, so the coop should be near the manure pile. The trees can provide forage for chickens, so their pen could open out into the trees. Lay the cards out next to each other on the table, so you can visualize these connections.

If you can't leave the cards undisturbed as you go on to the next phase of design, draw a flowchart of how the elements are organized and connected.

Working with Cycles

Many of the cycles of nature will continue around you without your notice, even as you become more attuned to your gardens and the various creatures that inhabit them. However, some awareness is necessary. The water cycle is the most obvious, with the wind and rain reminding us frequently of its processes. The soil goes through cycles that are evidenced by the plants that grow—the pioneer species fixing the nitrogen in the soil and preparing the way for other plants. Then there is the nitrogen cycle itself, which is just as crucial as the water cycle for life on Earth. The atmosphere is ripe with nitrogen, but plants are unable to use most of it. It must be chemically altered by bacteria (like rhizobium) or by lightning. Once it changes, it is *fixed*. The volume of fixed nitrogen in the soil decides how much will grow, and when those plants die, they release the nitrogen. There are also cycles of harmony between species. The ducks forage in the gardens in the fall after the harvest, which fertilizes the soil, removes pests and weeds, and feeds the ducks. They live and grow, providing eggs and meat for people.

The Ideal Farm System

When your garden is in its best working order, the soil will be completely mulched and the soil rich with organic matter. When you harvest a plant, as much of it will be eaten as possible and the rest composted in the compost bin, while you replant as many times as possible. Some of your dill, fennel, and carrots are left to *go to seed* (allowed to produce flowers and seeds rather than being harvested) in order to attract parasitic wasps. In the winter, a green manure crop is grown and turned over to return precious nutrients to the soil. Any volunteer tomatoes and cucumbers from the compost heap are replanted in an empty spot.

In a permaculture system, beds often take on a spiral, wild shape that flows with the contours of the land. This is where we need to talk about scale. There is an efficiency of scale that tips when we talk about the difference between urban farming and urban homesteading. Urban homesteaders should absolutely follow this permaculture ideal. It is easiest and requires less work. It becomes inefficient when you increase the size from several hundred square feet to several thousand square feet. A farmer interested in intensive production should have a standardized bed size and shape. This makes harvesting and replanting easier when you are planting on a weekly basis throughout the season. By the same token, in some cases bigger is not always better, which we talked about in the previous section.

If you have strong winds and need to protect the garden, you can build a barrier instantly with tires, which also act as a thermal mass that absorbs heat. The ground should be prepared first with newspaper and mulch. Then the tires can be stacked and filled with earth, compost, hay or whatever scraps you may have. At the top you can plant a species that is wind tolerant. This is all less than beautiful—straight, row-cropped beds rimmed by tires and plastic tunnels. But it is a proven system that produces well in almost any climate.

A Word about Animal Housing

You can look at the animal chapter to see the individual needs of animals, but in an urban setting you often can't be picky about animal location. Animals aren't sheltered in barns but rather in sheds, shelters, and pens. Here are a few tips:

- Orient the shelter at a forty-five degree angle to the prevailing wind so that it won't act as a wind tunnel. If this is not possible, build entrances from two different directions.
- Put it close to the manure pile and have it opening into the animal run. Water and feed should be nearby and it should be easy to remove manure.
- Use natural lighting as much as possible, and if you need lights make sure each pen has its own lightbulb with its own switch.
- The feed room should be rodent-proofed with sheet metal. Mucking equipment should be kept somewhere other than the feed room to prevent contamination.
- The floor is probably the dirt or grass of your yard. A thick covering of hay or sawdust is the solution to keeping animals' feet clean and making cleanup easier.

CLIMATE/ ORIENTATION

"For the man sound in body and serene of mind there is no such thing as bad weather; every sky has its beauty, and storms which whip the blood do but make it pulse more vigorously."

—*George Gissing, "Winter,"* The Private Papers of Henry Ryecroft, *1903*

Climate as an Asset

Weather is one of the top complaints of all gardeners: too much sun, too little sun, too much wind, and too much rain. Conditions are never perfect. North America has almost every climate represented, from the driest places in the world to full tropical rainforests, which not only increases the difficulty in growing plants but also provides tremendous opportunity to grow just about anything.

Most people cut a square out of their backyard and plant perfect rows of the same varieties of species that you might find at a grocery store, which is a very narrow view of gardening. Not surprisingly, they often meet with failure and give up. Not only are those species developed for a specific type of large-scale commercial farming, the chance that those species were designed for your climate and weather is very slim.

The real extremes of weather and climate can be used and defended with a little planning. The importance of choosing the right species is emphasized throughout this book, and it starts with climate. A hardy species developed specifically for your climate is likely to do better, obviously. You must set up microclimates using trellises and water diversion, either to retain moisture and coolness in the desert, or to reflect heat and store water in a colder climate. The key to this is remembering

that weather isn't a problem; it is an asset to be used.

Climate Zones

Most people are familiar with the USDA Plant Hardiness Zone Map. These zones are useful in a general sense, but in that case we might as well consider our land in relation to the whole climate, rather than through plant hardiness. There are three general climate zones (at least, three that humans live in):

Temperate: The *temperate* zone is where most people in North America live. In the winter the temperature drops below freezing, and in the summer it gets hotter than 50°F (10°C). There is a polar zone as well, but since it never gets warmer than 50°F, not much grows.

Subtropical: In North America, there are very few true tropical areas—at the bottom tip of Florida and some parts of Mexico. Instead, the humid, warm climates are *subtropical*, with winter temperatures that never drop below freezing. Unlike the tropics, temperatures do drop below 64°F (18°C).

Arid: There are many arid regions of North America, with an average rainfall of less than twenty inches (50 cm), or even desert with rainfall under ten inches (25 cm).

These zones are also at the mercy of the weather, and with the very real effects of climate change, people will experience greater extremes. Rain will fall more and create flooding, or rain will fall less and cause droughts. Temperatures will break records for heat or cold, and wind will become violent. More frequent deadly storms will occur. This means that a humid, subtropical climate may have even more hurricanes, or maybe the normally humid

weather will be uncharacteristically dry. Even if these changes are subtle, they have very far-reaching effects on ecosystems and plant growth. What may have worked in a garden for many years may not be as successful in the future. This is why plant hardiness zones are somewhat useless. Being flexible and smart in creating microclimates is a better strategy.

Rain

Rain may be one of the most important sources of water that you have, but it's not the only source of precipitation. Snow and hail fall from the sky, and condensation or *dew* collects on the earth. Water is a recurring subject in this book because it is the single most critical resource necessary for survival.

1. Awareness of the yearly average rainfall is important, but what is even more useful is being familiar with where it rains the most and how much of it you will get at one time. If all the rain happens in the spring and none in the summer, you will have to plan accordingly.

2. Trees planted above the house on a slope shouldn't need much water and should be able to thrive independently. Their purpose is to stop erosion due to rainfall.

3. Too much rainfall also limits the amount of sun. Cloudy skies block the light, and this is much more difficult to counteract than dry conditions that can be solved through adequate water storage. Reflective plants and surfaces can be used to increase the sunlight that falls on sun-loving plants like tomatoes or peppers so they will ripen.

4. Dew is most easily collected in the desert where there are clear skies and

a gentle breeze. You can stack some stones, which will collect moisture for the ground below. Several stacks surrounding a plant can keep a grapevine or small tree alive.

Frost

Frost happens when the land loses heat rapidly, usually just before the sun comes up. You need to understand your region's yearly frost dates and the lowest and highest temperatures the plants and animals will have to tolerate. Mulching the soil can prevent heat from being lost, and the frost will settle on top of the mulch rather than on the soil. But if there are plants coming out of the mulch, this can also cause them to freeze. The best way to prevent frost is to stop water vapor from condensing by keeping the air moving. Try not to create any blockage that would cause cold air to pool in one spot. In the southern hemisphere the south side of a slope is usually colder, and in the northern hemisphere the north side is usually colder. Situate your gardens on the other side.

1. Some plants require frost to produce fruit, and frost also adds water to the soil. It's just not possible to stop frost everywhere on your land, nor would you want to. If you are situated on a slope or hill, you can locate the gardens above the frost line (an area known as a *thermal belt*). This is important because cold air settles on the tops of hills, causing frost, and also flows downward, pooling in valleys. This leaves the thermal belt a little warmer.

2. A wall built with dark stones absorbs heat and can prevent frost by radiating heat at night. Plants will grow faster next to a wall like this, but a light-colored wall will assist some sun-loving plants to ripen, so use the color that works best for plants that will be growing there.

3. A large body of water will warm and cool more slowly than the surrounding environment, and thus serves to modify the temperature of its immediate vicinity. This creates a small microclimate, and this is why there is less frost near the ocean.

Wind

Too much wind can harm wind-sensitive plants, blow away seeds, lower the temperature of the soil, dry up moisture, kill young animals, and make it intolerable to live and work. Winds of fifteen miles per hour, which in some places is the average wind speed, are strong enough to reduce production. Winds of twenty miles per hour are enough to cause physical damage to plants. A windbreak can either block or channel winds in the direction you want them to go, ensure that plants produce and animals gain weight, be made of almost anything, be edible for animals and bees, and provide a home for beneficial birds. Near the ocean, a windbreak is the first priority in establishing an orchard.

When planning your sectors, take note of the direction of the prevailing wind, or the direction the wind blows more often. You do this by tying a streamer or flag on a tree or stake. In the city, the wind may not necessarily be a problem because you can use buildings. As mentioned before, you can use tires for a quick solution or potted trees for a natural windbreak.

Since cool air always goes downhill, a slope can create a wind that sweeps through a valley. In some large valleys, the wind will flow uphill during the day and downhill at night. The side of a hill with the prevailing wind will experience faster wind speeds going uphill, and as the wind reaches the other side it will be disrupted and slow down as it goes down again. A similar effect happens near large bodies of water: during the day, warm air rising creates a breeze that circles toward the land, and at night, as the air cools, it reverses and circles toward the water.

A small shelter, such as an old bag wrapped around stakes, a metal drum, old tires, or straw bales, can be built around plants to shelter them from wind. Traditional structures like *cold frames* (wood frames with glass tops), *cloches* (capes or bells that sit over the plant), and milk jugs with the bottoms cut off can all work. A building can be protected with bushes and vines planted around it. Even just snow or dirt piled up can be an effective insulator.

When choosing the material of your windbreak, remember that trees are both a benefit and a competitor. They provide firewood, stop erosion, create privacy, and offer a habitat for animals. They also have large roots that need lots of water, and windbreak trees won't give you as much fruit.

Another way to shelter the garden from wind is to build a trellis extending from the corners of the house, and plant climbing vines that grow up and cover them. These will control the wind flow, trap the sun, and provide something to eat. These grow fast enough to provide wind shelter as trees are growing, but make sure to pick the right species as they can quickly take over and embed themselves into structures permanently.

Temperature

Every 330 feet (100 m) of altitude away from sea level is equivalent to a temperature change of one degree of latitude from the equator. For example, if you are hiking exactly on the equator, and you climb to a height of 1,650 feet, the temperature will be the same as it would be five degrees from the equator. This is why a tropical region can have snow in the mountains. A body of water also modifies the temperature of the surrounding air through evaporation. In very hot areas, even a small body of water or a fountain can cool the surrounding area. There are various means of modifying the temperature:

1. The most obvious way to create a warm microclimate is with a greenhouse, which is particularly valuable in the winter.

2. In the spring, remove mulch from the soil in your intensive growing areas so that the dirt can warm up, because mulch doesn't have any heat-conducting properties. Mulch (and living ground cover) should be used at almost every other time in the gardens to help the soil retain moisture, reduce erosion, maintain a stable temperature, and stop weeds.

3. Plants release water vapor, cooling the air and causing humidity. Filling a porous earthenware pot with water and covering the top with a heavy, wet fabric can create a similar effect. Place these around in the area you want to cool.

4. Blocking the sun is a very quick way of cooling something down. Shade cloth that allows some light in is the simplest way of cooling the growing area.

Light

Plants need light, and redirecting sunlight can mean the difference between green tomatoes and red ones. Any plants that have a high need for light and warmth can be placed on the sunny side of buildings and slopes.

Trees with light-colored leaves can be used to reflect heat and light. A sun-facing wall can be used in a similar way, and can reflect the sun in winter. Plants will ripen faster and more completely if they have a reflective surface behind them.

A temporary shelter made out of fabric or other material can be used to provide shade and prevent sunburn during the hottest parts of summer.

WATER

"We never know the worth of water till the well is dry."

—*Thomas Fuller,* Gnomologia, *1732*

Assessing the Water Situation

Water is the most important element. On a piece of land, no other factor impacts so many other things, or is affected by so many other elements. How much water you have depends on the rainfall, how the soil

---

Beginning dates of seasons (for the northern hemisphere—they are reversed in the southern hemisphere):

Summer solstice: On June 21, the North Pole leans most toward the sun, and it is the longest day of the year.

Winter solstice: On December 22, the South Pole leans most toward the sun, and it is the shortest day of the year.

Vernal (spring) equinox: On March 21, the earth is most sideways to the sun; day and night are the same length.

Autumnal equinox: On September 23, the earth is most sideways to the sun; day and night are the same length.

Lengths (for northern hemisphere):

Spring: 92 days and 20 hr.
Summer: 93 days and 14 hr.
Autumn: 89 days and 19 hr.
Winter: 89 days and 1 hr.

▲ These rice terraces in Bali illustrate effective water management. The torrential rainfall running downhill is utilized by the water-loving rice.

---

drains, the plants that are currently growing, people and animals using the water, and what kinds of plants you want to grow. Some of these factors are in your control. The first step is to decide where the water is coming from and devise a way to store it, using gravity to move it to where you want to use it. The second step is to use species that need less water in places that are difficult to get water to. In an urban setting, this may be as simple as plugging a hose into a faucet, but awareness is still important.

While trying to reach these goals, keep in mind that water is also a responsibility. Your job is to use the water you harvest for as many tasks as you can through conservation and reuse.

Swales

A *swale* is a long, shallow ditch about three to five feet across that serves to stop and channel the flow of water into the soil. Unlike a regular ditch, which directs and carries water somewhere else, the swale is made to help the water absorb directly into the soil. Swales lie across the land, especially across a slope, and when the water is forced into the soil, it can be soaked up by trees planted along behind. They can be filled with rock, gravel, or gypsum for even better water penetration. Swales are perhaps the most effective method of water conservation in both dry and humid climates. They work well on steep slopes or on flat prairies and can even be implemented in an urban area to take advantage of road and roof runoff.

A swale may not be at all practical for the urban farmer, because it requires pretty drastic changes to the landscape. However, in this time of climate change, knowledge of swales will be increasingly important. It may become the difference between having and not having food, and so it is included here.

Building a Swale

It is not difficult to dig a ditch a few feet deep. The swale will catch the water in a pool as the rain runs into it, where it will gradually soak in. If the swale is overflowing, then you need to widen it or improve the drainage. Over time you can throw mulch in and grass will probably grow. If you plant trees along it, they will get taller and begin to shade it, and other species will spring up.

The swale is really all about positioning. Two or more are always put in together, along the counter of the slope (horizontally crossing the slope to stop the water from running downhill), and the soil is loosened up to help water penetrate. The second swale is placed between ten and sixty feet away from the first, depending on the amount of rainfall you get per year.

Rainfall	Space between swales
Less than 10 inches	60 feet
10 inches	50 feet
20 inches	40 feet
30 inches	30 feet
40 inches	20 feet
50 inches	10 feet

Drains

Unlike a swale, which stops the water and forces it into the soil, a *diversion drain* is a ditch that carries water away. A diversion drain can be used to direct water into a swale, a pond, or an irrigation system. Drains are also used to direct the flow of water to a place to catch water runoff. If they are directed to swales, they don't need to be waterproof, but if you are sending the water to a dam, the drain should be built of rock or concrete. Installing a *spill gate* also gives you control over the flow of water. If you fall victim to flooding or you want to control crop irrigation, you can use the spill gate to direct the water where you want. A spill gate is simply a removable method of blocking water.

Dry Bed Management

This kind of management is most valuable in the desert where small creeks and streams can quickly get out of hand after a short rainfall. The rain runs into the low creek beds and rapidly becomes a flood, eroding the soil and becoming unusable to anyone. It flows too fast to be absorbed into the soil or taken up by plants that conglomerate in the creek bed. It may also cause damage.

Stream *braiding* is a nontechnical term for spreading the flow of a stream out into a myriad of much smaller channels over a landscape, so that it irrigates the entire area, and at the same time prevents flooding. This happens naturally in many areas where silt and sand erode and deposit downstream, breaking up the terrain. To do this, start at the head of the stream and dig a small pool. Branch off from that pool and create a diamond pattern across the landscape by digging diversion channels. Where each diamond intersects you will dig another small pool. This strategy works well for streams that tend to run dry parts of the year and are prone to flash floods.

Cisterns

A small farm under two acres can use between five hundred and one thousand gallons (1900–3800 L) of water per day during the summer, or more. Tanks can be made from plastic, concrete, compacted clay, metal, or even plastered dirt. The water to fill the tanks comes from rainwater that runs off roofs of buildings or other surfaces, or is pumped in from a well. To prevent mosquitoes, the tank should be covered and screened. Thick, green algae will begin to grow over the sides of the tank, but this is a good thing because the algae will help clean the water. The outlet pipe of the tank can be three inches (7.6 cm) above the bottom so that the algae remain undisturbed.

Some people have recommended that mosquitofish can be used in a tank to eat mosquito larvae. The use of mosquitofish (or *gambusia*) does more damage than good, and it is questionable if they eat any more larvae than any other fish. They have hundreds of fry and breed prolifically, but they only live a couple of years. They should not be introduced to your aquaculture pond, because other fish tend to avoid eating them and they can quickly choke out other populations. A more effective mosquito control is frogs and birds.

It makes sense to place a water tank at the top of a hill. In fact, a large water tank set on the top of a hill can act as the foundation for a building, and the building roof can be used to collect rainwater.

▲ This eight hundred gallon plastic tank was traded with a family for a CSA share. It served a one-and-a-half acre area. A pump ran on a timer through the night to fill it, and a second timer turned on the sprinkler.

Choosing a Tank

There are many different types of water tanks. Choosing the right type can be difficult because the pros and cons aren't always extremely clear. Price is always a factor. The earthbag tank is cheap and durable, but it takes up a huge amount of space. Polyethylene is cheap, durable, and comes in any size and shape, but people concerned about chemicals leeching from the plastic will want to avoid it. Even an earthbag tank is lined with polyethylene, so it may not be any cleaner. Galvanized steel is considered safe to drink from, although it has a zinc coating that does leech into the water. It is a good thing that the zinc no longer has lead in it, which manufacturers at one time used. The concrete tank is less expensive but it has more metal in it in the form of rebar than a galvanized steel tank does, and even then its lifespan is shorter than that of other options. However, if the right concrete is used, it may be the safest to drink from.

Another major factor is portability. If you believe that someday you may need to move your tank, plastic is the clear choice.

Earthbag Cistern

Earthbags make a very inexpensive water tank. You must use fifty-pound bags made of polypropylene or other durable plastic that will be waterproof, at least seventeen inches wide and thirty inches long. The steps are simple, although very labor-intensive:

1. The tank is set at least a quarter of the way into the ground, and so you have to dig a smooth hole the size and shape of your tank, leaving room for the earthbags.
2. Line the hole with pond liner or heavy polyethylene.
3. Begin building the walls. As you put down each layer of earthbags, lay two strands of barbed wire between each layer to hold it together.

Type	Pros	Cons
Galvanized steel	Durable Takes up little space Lasts 20 years	Must have a concrete base Can't be set in the ground More expensive Possible metals leeching
Polyethylene	Available in all shapes/sizes Less expensive Doesn't need a concrete base Durable Lasts 25 years Takes up little space	Possible chemical leeching from plastic
Concrete	Less expensive No chemical leeching Lasts 15 years	Has more metal in it than a steel tank (rebar) Takes up more space
Earthbag	May be cheapest Lasts 20+ years	Takes up more space

4. Once you have the walls done they need to be waterproofed. You can plaster walls with concrete or stucco, lay down a pond liner, or use a sheet of heavy polyethylene. Cover the top with an earthbag dome or concrete.

Cold Climates

In a cold climate, the water is likely to freeze over and make a certain amount of your water unusable. Several feet of ice can make a significant difference in your water supply. Pipes need to be buried in the ground at least three feet to prevent freezing and bursting, or they will have to be left on all winter or completely drained. Such wastefulness can also quickly drain your tank. You can prevent many of these issues by putting the water tank under the ground.

Pond Design

Water holds and reflects heat. As the temperature drops at night, heat radiates into the air and surrounding gardens. A pond will keep plants and people warmer

▲ A small backyard pond versus aquaculture tanks.

in the cold and cool them off in the heat, and serve as a place to grow water-loving plants and fish. Tiny frogs will live there and eat bugs. Your pond can grow water chestnuts, rice, bait fish, freshwater shrimp, snails, aquarium fish, water lilies, basketry materials such as reeds and rushes, and mushrooms. It can also be a home to crayfish, prawns, mussels, clams, and ducks.

Some rules of thumb for designing ponds:

- For fish, several small ponds no more than four to six feet deep work better, while storage ponds can be ten to twenty feet (3–6 m) deep.
- A large pond for irrigation shouldn't have many fruit or nut trees around the pond, or the pond will become polluted with leaves and dropped fruit.
- Any waterproof container can be used. You can use stones to disguise the edges of a bathtub sunk into the ground.
- Large livestock can't have access to ponds for aquaculture or they will destroy the balance of the pond.
- The pond must be near a water source to be filled initially or topped up during a dry season. This can be a diversion drain or hose.

The pond should have various refuge areas for each species that lives there. For the ducks, a small island can be formed in the center. A series of shallow shelves around the edges can serve as a home for forage plants. The pond also needs to have some deep areas at least six to eight feet deep (2-3 m) where fish can retreat in the summer when it is hot if the pond is outside. Drop in some hollow logs or pieces of pipe for them to hide in.

The depth or volume of the pond doesn't have any impact on how many fish you can stock. The population of the pond relates to the total square feet of its surface. A pond three to six square feet (1–2 sq. m) can be used to raise some plants and a small population of fish. A pond at least a hundred fifty square feet (14 sq. m) can supply a family with all the fish they need (if the right species is selected), water plants, and a flock of ducks.

Pond design isn't extremely complicated. As long as you use any swampy areas or low-lying ground you already have and waterproof the bottom, you can build a fairly successful pond. Keep in mind the rules of thumb to be even more successful at growing things in it.

Constructing the Pond

1. Lay down a rope as a guide for the shape of your pond while you dig.
2. Divide the pond into thirds: one-third for the shallow end, one-third for the mid-range depth, and one-third for the deep end.
3. Line the bottom with pond liner. *Pond liner* is a heavy black plastic, rubber, or geotextile sheet made specifically for this purpose. The edges of any small pieces should be taped together or weighed down with heavy rocks. You may also want to put down a layer of pond underlayment under the liner to protect it.
4. The edges of the pond will need to be stabilized with rocks of varying sizes, bamboo, grass ledges, or logs, especially at first, before other plants step in to prevent erosion. This step also helps hide the liner, so it's a good idea to use several materials for a natural appearance.

5. Do not introduce fish right away because there won't be enough food. Once you have built the dam and the pond begins to fill, lay down a couple of inches of straw and trample it into the bottom. This will provide a habitat for water insects.

6. Introduce water plants such as lilies, water chestnuts, and duckweed to build up the bug population. This process may take at least six months. They have to be very well established to withstand being eaten all the time by fish and other animals. You can start this process by leaving the plants in plastic pots for a while and simply submerging them in the water, so you can move them around if you need to. Transplant them when they get bigger.

7. The pond inlet from any outside water source must be planted with grass to help filter the incoming water. Keep the inlet clear of any debris.

8. The water should become green. If it doesn't, add a small amount of manure. The water may also be muddy at first if it is coming from a dam or other flowing source. If that is the case, add one teaspoon of gypsum per square foot (4.5 g per 0.10 sq. m).

9. If the water gets too warm or it has been cloudy for a while you may need to aerate it, or the oxygen in the water may drop too low. Plants won't oxygenate the water unless it is sunny, and they may not be able to keep up with the demand in hot weather. The average water temperature should be about three-quarters of the air temperature. This is why people have tiny waterfalls and pumps that keep the water flowing. However, we are trying to save energy, so you might want to invest in a floating emergency aeration system. That is an electric device that sits on the surface of the water and creates bubbles and disturbance.

10. Hopefully you won't need an emergency aeration device if you have the right proportion of plants. The plants should cover around 60 percent of the surface of the pond. Plants that live under the surface should be spread at a rate of one bunch per two feet (0.2 sq. m).

11. Introduce baby fish in the spring. You can also add a few buckets of pond water from a nearby pond to introduce a supply of aquatic insects.

12. Over time the bottom of the pond will acidify, and while you can add lime to balance it out, every few years you may need to drain it. When it is dry, you could raise a crop of melons or rice. Your primary goal is to maintain water quality, with a pH of around seven or eight.

Irrigation Principles

Even a small property can become very self-reliant in the water department with a well-developed water diversion system. It is not only ecological to use your own water sources from your own land, it is more secure. You will have peace of mind knowing where your water comes from and what's in it, and that you will never be without it due to some disaster. Rainwater that directs into swales and reserve ponds may have irrigation pipes leading to the garden and orchards, which also have swales and diversion drains helping to direct water to the right places. However, an irrigation system should never be used for growing watermelons in the desert, or for growing lawns and washing cars.

Irrigation is simply a way of supplementing the natural flow of the land, and possibly rehabilitating the soil.

Desert Irrigation

With some careful planning, you can still have enough water to grow food in the desert. All gray water should be recycled to a wetland marsh, and instead of spray or flood irrigation, a drip hose should be used. The drip hose should be buried at least six inches under mulch or soil. Drip hoses can be expensive, but you can make your own version with earthenware pots sunk into the ground (called *ollas*), bottles with holes punctured in the sides, or pipes filled with gravel. In the orchard a sprinkler can be used to spray a mist over roots. Sprinklers are only used in shady tree areas; otherwise evaporation becomes a problem. In dry, hot climates, only water in the early morning, late in the evening, or at night. Mulch and swales, which trap and preserve water runoff, are really the secret to successful desert gardening.

Designing an Irrigation System

Unlike "irrigation systems" that people shell out thousands of dollars for, permaculture irrigation is half species choice and half terrain. Only a small part of it has to do with piping or pumps or equipment. By studying the slope and topography of the land, gravity is used to direct water into the gardens or even to individual plants. These individual plants are chosen for their specific traits that make them function well in the location that you want to put them. The roof is harvested for rainwater, which is stored in a large tank and hopefully gravity-fed to the gardens below if it does get dry. It is less likely to dry out if the topography and layout of your

garden are conducive to water retention, if you have mulched extensively, and your species can handle a little drought if you can't water them.

Simple Rainwater Collection

The simplest system is one in which a roof or other sloping surface is equipped with a gutter that carries the water to a covered storage container with an outlet tap. The old term for this system is the "rain barrel."

Awesome Rainwater Collection

- 1,500-gallon cistern of nontoxic metal or plastic (or homemade cistern)
- ½ horsepower shallow-well pump
- Plastic PVC and CPVC piping

Every 1,000 square feet (92 sq. m) of roof surface area will gather 600 gallons (2,270 L) of water per inch (2.5 cm) of

rain, and every roof on your property can be used to collect rainwater. The best material for this purpose is metal because it's a bit cleaner than other roof materials. You will also need to install gutters with a leaf screen.

Making It Work:

1. The roof of the house is where the water is collected. The bigger the roof the more water you will get. The rain washes off the roof into the gutters. Metal roofs are best because they easily wash and collect the most water.
2. The gutter is covered with a leaf screen and directs the water to the cistern. A mosquito screen covers the entrance hole of the cistern, and an overflow pipe at the very top directs extra to the garden.
3. From the cistern, the water goes through a valve to the pump.
4. The water leaves the pump, and through your timer to a multiple-outlet irrigation valve. The timer will control which hose is on and when, and for how long. Today, technology exists that allows you to control this with your mobile phone and automatically check the weather.
5. In the city, if you still want to be connected to the municipal water supply (which you will need to do to have enough water for irrigation), you usually need an approved backflow prevention device that will ensure that no rainwater gets into the city water supply. This device must be examined and maintained properly.

Gray Water

Gray water is the water that drains from sinks, tubs, washing machines, and showers. If you use a separating compost toilet, then you might divert the liquid wastes from that as well. Gray water is different from *black water* in that it doesn't contain sewage: solid human waste. In most systems the gray water is simply mixed with black water and dumped into a septic tank or sent to the city sewer. This is obviously a waste. Gray water, if handled properly, reclaims valuable nutrients and puts them back into the soil, while at the same time saving energy and water.

Keep in mind that gray water systems might be illegal where you live, although it is changing. Where building codes do exist to regulate them, you will have to get the system inspected and approved.

Gray water cannot be spouted directly into a food garden because it will contaminate your food. It can, however, be piped into a marsh or container garden where it will help grow a variety of water-loving plants. The first consideration is location. Either you need to use an existing marsh, or you will need to pipe it somewhere where nothing important will be contaminated. This means a place that won't flood, isn't near any food, and where it won't harm any important species. To build your own gray water garden, follow these steps:

1. From the gray water source, install a two-inch (5 cm) pipe from the house to the gray-water treatment area with a grade of 2 percent, put through a trench on a bed of sand. Install a shutoff valve and two screen filters: one on the inlet and one on the outlet. Fill the trench with soil.

2. Calculate your water usage. A frugal household of five with a waterless composting toilet uses about 940 gallons (3,558 L) of water per week.

3. Your runoff area should be able to handle the volume of water that is entering it in a day. One cubic foot (0.03 cu m) of wetlands will filter roughly one gallon (4 L) of water. Your surface area should be able to handle about a third of your daily gray water production, and the container should be two feet deep (60 cm). Here's the formula:
   Daily gray water (in gallons or liters)/**3** = surface area of marsh
   Example:
   134 gallons per day/3 = 44 square feet

This could mean a long skinny marsh two feet wide and twenty-two feet long (0.6 m x 6.7 m), or a small area five feet wide and nine feet long (5 m x 2.7 m). But in an urban environment, an alternative method is used. Rather than building an open marsh, the pipes run into a container garden. This container garden would have the same capacity and organic materials as the marsh, but would have to be formed of either one or more concrete planters.

4. Build your concrete containers. The inside of the planting container should be layered with a variety of filtering materials, starting with a layer of gravel, then sand, then a layer of sticks cut into six-inch (15 cm) lengths, and finally a layer of straw and other mulches. The mulch must be at least eight inches (20 cm) deep and will have to be replenished every year.

5. Now you can begin growing wetland plants in the mulch. Local varieties of cattails, rushes, reed grasses, horsetails and water-loving flowers are good choices. The cattails are edible, and rushes and reeds can be used for making baskets. These plants are part of the filtering system, purifying the water as they take it up into their roots.

6. There are a few rules to using gray water. You must use biodegradable soaps and avoid any detergent. You also cannot wash anything with human waste on it, such as cloth baby diapers, and send the water to the gray-water system. If you have diapers to wash, you will have to have a way of keeping this black water separate. You could have a valve for the washing machine that sends water to the city or septic if you need it, and then switches back for

other loads. Or you could give them a rinse and wash by hand and dump the water in your humanure compost before putting them in the washing machine. Having a *humanure* compost pile is questionable in the city, however.

7.   The water level in your containers should never flood over the top of the mulch. If you find that you are over capacity, you can install a tank with a valve to control the flow of water, or you will have to increase the size of your system. If you do use a tank, the liquid in it must cycle every twenty-four hours or it will be too toxic to use anymore. Watch for clogs, and prevent them by flushing the system with clean water once a month.

This system can be valuable for the urban farmer for two reasons: First, it can provide a cash crop of ornamental water plants. It can also provide you with a sheltering plant system in case of drought. Drought is going to be a major problem in our changing climate, and shade and mulch will be an essential part of growing food. The plants can be used as a mulch. The other alternative in a permaculture system is to grow plants near the edges so that they shade the soil, preventing evaporation.

Ways to Water

Bucket: Many people use a bucket or watering can with great results, if they are watering inside or in their own backyard. It is the cheapest, most tiring method, but quite effective and it doesn't waste much water. Watering with this method is very time-consuming, and just not possible if you are using an area bigger than a few hundred square feet.

Upside-down bottle: Used mainly for trees, get a jar or jug with a small opening and fill it with water. Stick it in the ground upside down next to the plant.

Drip hose: A hose with small holes that lets water drip out directly onto the soil and next to each plant. The cheapest version of this is called T-tape, although it degrades over time. Your local irrigation supplier or garden store can tell you what you need and how to install it.

Soaker hose: A hose with small holes on one side and a cap on one end that forces all the water out like a sprinkler three to five feet.

Ditch irrigation: A ditch around a field through which water from a well, spring, or community irrigation ditch flows. It is diverted into the field by contour ditches six to twelve inches deep through the field downhill. You don't want to wash things out, but you don't want it too dry. To make the water go into the contour ditches, dam the big ditch.

Traditional irrigation: Big farms use water pressure and large rotating sprinklers, either with wheels or simply rotating heads. These are quick and easy, but expensive and wasteful.

For urban farming you will be using everything but traditional irrigation. For things like microgreens, a pesticide sprayer that has a fine mist works well because it doesn't disturb the seeds like a bucket would. A drip hose is your most likely solution, with a timer. For plants in containers where a drip hose won't reach, you will probably be using a watering can or bucket. Watering will take almost as much time as weeding, so plan your irrigation carefully.

Summing It All Up

The basic rule of watering a garden is to put enough water to reach the roots, and when it dries out, water again. Generally, this means that you need it to be moist about six inches down (you can dig to find out), and you will have to water in the morning and evening (unless you have wet weather). Soak them good and deep when the sprouts are well started. If you are using mulches and shading the soil, you won't have to water as often or as much.

In the coming years, drought and flood will make farming much more difficult. How can you keep the soil damp all the time during a water shortage? How can you drain a marshy backyard? This is why we have talked so much about rainwater collection, water storage, and redirection of water. Mulch is the key to the future of food production. It is used in *dry farming*, which is just creating a sponge out of mulch to absorb rainwater so it can be used during dry spells. Mulch also prevents erosion during flooding or high winds.

FINAL DESIGN

"It is not the strongest of the species that survives, nor the most intelligent that survives. It is the one that is the most adaptable to change."

—Charles Darwin

The Successful System

- High biomass production. *Biomass* is the volume of living organisms in an area.
- Lots of organic matter. Even if it's not alive, the organic matter in and on the soil should be tremendous.
- Living organisms provide minerals, rather than rocks and rain.

- Retains minerals over time. Consumption and erosion have little impact.
- Allows fungus and bacteria to play a central role in the cycles of life.
- Maximizes diversity.
- Has complex food chains rather than simple ones.
- Sees relative sameness over time. Very little or very gradual change.
- Uses species that are useful in as many ways as possible.

Drawing It Out

You've already taken an inventory of your land using permaculture principles. You should know how many square feet your garden space is, which way the ground slopes, the direction of the sun (or light source) and where it hits, and where there is shade. You should also know how far away and what direction the water source is and use graph paper to draw a picture of all these things. But you still can't draw it out until you have decided on your plants, varieties, and your calendar. You probably have a general idea of what you want to grow. If you are following a typical urban SPIN-type model, you'll be following a variation of Curtis Stone's fifteen or so vegetables. If you are doing farm-to-table, you might have a garden that specializes in herbs or gourmet mushrooms. Your profitability is based on your square footage and how much you can grow per square foot. It is time to map out your space, what you will grow there, and the other elements.

Trellises

The house, the fence, the outbuildings, the shaderoom, the patio—these are all possible *trellises*. A trellis is simply a

framework that supports plants. Trellises can be built over pathways, and in the middle of gardens in the form of tripods, circular baskets, wire frames, archways, or even over streams. They can create hedges if perennials (such as hops) are planted on them, they keep areas cool, and they become part of living spaces. Ivy, grape, wisteria, roses, beans, cucumbers, melons, squash, peas, and tomatoes all need trellising.

Cold Frames and Covers

Unlike a high tunnel, which is more or less an immobile greenhouse, cold frames and covers can be used over any bed. This is very valuable in high-rotation beds. You can throw thin plastic over your plants and hold it down with rocks, or you can use a cloche, which is a light frame holding the plastic in a dome shape. A cold frame is a wooden box with a glass lid that can be opened and closed. To use a cold frame, you simply cover the plants when you know there will be frost. Generally, this means close them at night and open them during the day. Cold frames are harder to build and difficult to move but last longer and protect your plants from the weather. Cloches and plastic are easy to make but get wrecked easily. An easy cold frame is to use straw bales to hold up an old window—be resourceful. To make a sturdier cold frame, build a six-by-twelve-foot box on the ground with no bottom, and face it south. Angle the top so that the higher end is on the northern side. Fasten two windows with hinges on the top as a lid. Use a stick to prop the windows open. This is like a miniature greenhouse.

If you are extending your season by planting several gardens during the year, the early and late gardens will need special care. Some plants are hardy to frost, others aren't. For a garden planted in August you will especially need frost covers over the plants. This means that you will have to be extra vigilant in watching the weather and be dedicated to covering them up.

The Greenhouse

A greenhouse is a structure that allows heat and light from the sun to penetrate but doesn't allow all the heat to escape. Greenhouses are used mainly to grow plants that need the warmth, to extend the growing season, and to protect from frost. They can also be used as an attachment to your home so you can utilize solar heat.

The easiest greenhouse is a frame made of plywood or PVC piping that is fastened firmly to the ground, which is then covered in heavy-duty plastic. The bottom edge can be a wood frame or a square made of bales of hay, and the plastic

stapled to it. The plastic can be fastened to the frame with leftover one-inch PVC piping cut into two-inch lengths and sawed along the edge to form a "C." This is then snapped over the pipe and the plastic.

A more expensive and difficult greenhouse is a wood frame that is covered in glass or Plexiglas or other hard plastic sheeting. The frame would be similar to a barn frame, with gussets (or reinforcements to the joints) to strengthen it.

Fencing

Fencing is sometimes a major challenge for urban farms, which might seem ironic in a place where there are so many fences. You may be lucky enough to be in a location which already has fences, but if not, you will have to decide whether

▲ This high tunnel was built with PVC pipe and heavy-duty plastic sheets taped with Tuck tape. The wood framing on each end and along the bottom gives it strength.

or not to invest in permanent or temporary fencing. Obviously, the more permanent your property is the more permanent your fencing can be. But if temporary fencing is your only option, there are things you can do. Deer are usually a pest in the city, and you can use deer netting. This should be eight feet high and held in place by rebar stakes. Leave at least six inches of material at the bottom, which should be buried in a trench so they can't get under it.

My personal experience with deer is that deer netting will not deter a determined buck. After the buck jumped over and tore through our eight-foot deer netting, we used heavy plastic sheets to construct a fourteen-foot barrier. However, this buck also tore through the heavy plastic and jumped over nine-foot sections, and led four or five doe into the garden. This number of deer will decimate an entire season's growth in a single night. In that situation, you only have three options: find a bow hunter and eat venison, let the deer have it, or buy metal fencing.

3 | Methods

"Crabgrass can grow on bowling balls in airless rooms, and there is no known way to kill it that does not involve nuclear weapons."

—*Dave Barry*

In permaculture, a suburban area is not usually considered the most ideal growing space. This means that you will have to grow things very intensively. But even those who don't have a yard and are working with just a patio or windowsills can grow a surprising amount of food in containers.

Root crops like potatoes need to be grown in a deep metal drum, wood box, or stacked tires, in a bed of mulch and covered with more mulch. As they grow, put more mulch in around the stem so that it is forced to grow up out of the barrel. As you fill the barrel with mulch it will grow higher to reach toward the light, and this will force it to produce more potatoes.

Super-nutritious varieties that work particularly well for container gardening are pepper, parsley, tomato, chive, Swiss chard, and lettuce.

If you have very little space, stick with herbs. Another way to maximize your space is to build upwards. Hang baskets, build a greenhouse that sticks out of the window, and build shelving up the walls. Of course, it's always a good idea to be growing sprouts in the kitchen and a sack of mushrooms in a dark place.

If you do have room for a fruit tree, it should be a dwarf variety. If you already have a fruit tree, keep it pruned back so that it doesn't shade your garden. When you plant a new tree, you can plant it against a wall or fence and train it to grow more flat. It will have to be tied and pruned but it will save an amazing amount of space.

There are tremendous resources available in the city because of your close proximity to people. You will have to bring in organic material, but you should be able to find it where it would otherwise go to waste. City maintenance for the parks, construction projects, neighbors, and Internet communities like Freecycle can all offer sources of compost, wood brush, grass clippings, and other valuable "garbage" that you can use. Your neighbors are likely to be growing plants, too, and bartering can offer more variety to your diet.

One concern with city food production is lead. Lead can come from paint that was used on a house, or even dumped by some irresponsible person decades ago, and it can come from the exhaust of nearby cars on a busy road. If you suspect that there may be lead in the soil, get it tested. The solution is to add as much organic material to the soil as possible, which will make the lead less available and thus less likely to be absorbed by a plant. Be aware: If you gather organic materials from the city, avoid collecting leaves that are right next to a busy highway. Choose leaves that are on quiet side streets.

SOIL

A Little Bit about Soil

What is the difference between *soil* and *dirt*? Generally, most people think dirt is a dead substance that is not capable of growing a living thing. The dust of the desert or land cleared for building a house is just dirt. They may also think that soil is the stuff you buy in bags decorated with smiling cartoon flowers, or that it comes in trucks from the local supply store. In other words, there is an idea floating around that humans make soil, and the earth makes dirt.

Nothing could be further from the truth. Except in the case of some very big deserts, most places in the world had very fertile soil at some point. Infinite communities of bacteria and microbes create a complex layer cake of matter, gases, and liquids that is perfectly suited to the root systems of plants, and this happens all on its own when left undisturbed. It is a complex cycle that does quite well without our interference. When humans arrive on the scene and set up farming and civilization, we create dirt. Dirt is soil that has been displaced, with orphan particles missing key pieces that make it alive. We

▲ This soil contained a lot of bark and small rocks, but because of the mixture of compost and nitrogen it produced a healthy onion crop.

do this because we are ignorant of the microscopic life living below the surface of the ground. We destroy the soil's innate ability to regulate the flow of water, which then stops it from adequately filtering and cleaning the water, and in the process we rob it of the power to store and provide nutrients.

This, of course, creates all kinds of monumental problems. Erosion, crop loss, flooding, dust storms, dropping water tables, radon, salt buildup, and even allergies and asthma are directly related to humans' destruction of soil. Dr. Walter Clay Lowdermilk was a noted soil conservationist who worked as the assistant chief of the USDA in the 1930s. He travelled extensively all over the world studying the soil, and at the end of his tour he gave a series of lectures that became a pamphlet in 1948 titled *Conquest of the Land through Seven Thousand Years.* He documented the causes of war, the decline of civilizations, and the destruction of society. In his opinion, it all came down to one thing: "For even you and I will sell our liberty and more for food, when driven to this tragic choice. There is no substitute for food." Food will not grow without soil.

Soil Is Alive

What we call soil is really just a container for a microcosm of living creatures, most of which we can't see. A single shovelful of soil contains more species than are found in the entire Amazon rain forest. These are some of the organisms present in soil:

Bacteria: These convert organic matter into something plants can use and sometimes grow to have beneficial relationships with the plants directly.

Fungi: We usually know fungi as mushrooms, but the important part of a fungus is its system of rootlike threads running through the soil. These threads decompose matter just as bacteria does, but they also build soil nutrients in partnership with plants. Some might kill plants, but many others actually prevent diseases.

Protozoa: These organisms eat bacteria and keep the helpful bacteria releasing the nitrogen that plants need.

Nematodes: Nematodes do some of the same things that protozoa do, and they also travel around spreading live bacteria and fungi through the soil. Some of them prevent diseases and provide food for helpful predators.

Arthropods: These are insects that you can sometimes see, but often can't. Ants, beetles, sow bugs, spiders, mites, centipedes, and more all live together in the soil, chopping up materials and eating other bugs and fungus. They simultaneously mix the soil and make organic material more accessible to plants.

Worms: Worms are the powerhouses. They eat the organic matter in the soil and their *castings* (droppings) are perfect plant food. Worms create thousands and thousands of pounds of casts per acre.

Organic Matter

All those living creatures feed off each other, but they need one thing to get started. Organic matter is at the bottom of their food chain, the fuel that keeps everything going. *Organic* in this sense means organisms: living and dead creatures and plants. When any kind of organic matter is added to the soil, all of those living creatures get to work breaking it down, which adds nutrients to the soil that plants need. Once the matter breaks down completely, it becomes humus. *Humus* is organic matter that has decomposed so fully that whatever is left will not break down any further. Besides being the catalyst for all the biological processes that occur in the soil, organic matter is the stuff that holds water and prevents *runoff*. Runoff is exactly what it sounds like—rain or snow falls on the earth but can't be absorbed, so it just runs off the surface. Without humus, the soil erodes and loses nutrients, leaving plants to die of thirst and starvation. Soil organisms are able to decompose organic matter at such a fast rate that continuing perpetually requires massive amounts of organic matter to maintain the same nutrient content and life cycle. In nature, the plants provide the input by dying and falling right back into the soil, but in farming this material is taken away and consumed by people. For example, according to the Natural Resources Conservation Service (NRCS), an acre of wheat must have at least two to three tons of organic matter or more added to it every year to maintain the same quality. That same wheat field has thirty thousand miles (24,140 km) of roots just sucking up all those nutrients.

The Nitrogen Cycle

Life on Earth hangs in a careful balance. All organisms require nitrogen to live, and it is even part of our DNA. Most of the nitrogen available to us is found in the atmosphere, but it first has to be converted to a different chemical form for any organism to be able to use it. One of the single most important bacteria we know of today is rhizobium, which lives in a symbiotic relationship with only some plants. It exists in nodules in the roots of

legumes like beans and peas, and *fixes* the nitrogen found in the soil so the plants in the vicinity can use it. Fixing it really means converting it into ammonium. Some other things can do this, too, such as blue-green algae, lightning, forest fires, and molten lava. Once the plants absorb the ammonium, they eventually die and decomposers move in. Most of the now-unusable nitrogen contained in the plant is converted back into ammonium again and released into the soil where other plants can use it, which is the optimum scenario. However, some of this ammonium doesn't make it to plants because other soil bacteria convert it into nitrates. Unlike ammonium, which binds to the soil, nitrates are loose and just wash out with water. This leads to a loss of soil fertility and creates a nitrate buildup in streams and groundwater. Some of this nitrate also turns into gas, which goes into the atmosphere to become smog and greenhouse gas. Once it does that, it will never be part of the nitrogen cycle again, and is essentially lost.

Humans have more than doubled what would have been a relatively steady global nitrogen-fixing rate without our interference. By burning fossil fuels, growing mass quantities of legumes like soybeans, and through the use of synthetic fertilizer, we have pushed the nitrogen cycle beyond a sustainable level. Synthetic fertilizer was invented in the early 1900s and involves high temperatures and pressure to fix nitrogen, in much the same way that a cataclysmic natural event would. Chemical fertilizer makes up the vast majority of the fixed nitrogen we create, and while it can be used to grow more crops, it has disastrous consequences. While organic ammonium does leech nitrates into our drinking water, it is at a very low level that we barely notice. When chemical fertilizer leeches out, however, it creates extremely high and dangerous levels of nitrates in our drinking water. Nitrates lead to cancer and breathing problems for babies; they kill fish and change the biological balance of fragile coastal ecosystems. When emitted into the atmosphere, they become acid rain. The careful balance of the nitrogen cycle, which we depend on so completely for our survival, is in extreme danger. This may be the biggest argument against chemical-based farming.

Measuring Soil Health

Besides merely looking at the structure of the soil to see how much humus is there, we can measure the relative health of the soil by testing its level of acidity and alkalinity. If you remember high school chemistry, soil testing uses the pH scale of 0 to 14, with low numbers being more acid and high numbers being more alkaline. A pH of 7 is neutral, and most plants like 6 or 6.5. A soil testing kit can be bought from any garden supply store, and you will find that your soil will usually range from 5 (very acid) to 8 (medium alkalinity).

Soils that are very acidic (with a lower pH) can't hold onto nitrogen well. The lower the pH, the more chemical nitrogen fertilizer will leech away. This affects natural nitrogen processes as well—rhizobium just doesn't survive in highly acidic soil.

Plants need other minerals as well, and a good soil testing kit will include tests for them. The solutions to these soil problems are not complex and don't require any kind of chemicals. If the soil is too acidic, you can add agricultural lime (a mineral) or wood ash, which is sprinkled on the soil at a rate of twenty pounds per one thousand square feet, a little more if the soil has a more claylike consistency. Doing this also increases the potassium (K) levels, which we will talk about later. Other minerals like sulfur and trace elements like manganese tend to decrease as the soil becomes more acidic, and the most effective way to increase them is with compost, which also decreases soil acidity. Compost and peat moss will raise the alkalinity and help with a host of other soil issues. As soil today becomes overused and overfertilized, it loses its ability to hold onto the nutrients that plants need. Compost is the solution to most of these issues.

Over time, any soil can be changed and improved. This means that you could theoretically put your garden anywhere you want, although scoping out where the best soil is makes more sense. Armed with a little knowledge about the soil, you can grow anything.

Next you'll need to find out the drainage capacity of the soil. To do this, dig a hole one foot deep (30 cm). It doesn't need to be a wide hole. Fill the hole with water, and after five minutes fill it again. See how long it takes for it to drain completely. For some people, this might be never. For others the water might disappear immediately. If it takes more than four hours to sink in, you have a problem.

Look at the plants that are already growing. Bare soil is damaged by the sun, wind, and water, which is why permaculture insists on leaving soil undisturbed. Usually a plant that has become overgrown in an area, as is often the case with blackberries, indicates that the soil has been damaged. These plants act as pioneers and prepare the soil for the next stage.

What are the parts of the soil?

Humus: Organic matter in the final stage of decomposition, unrecognizable as plant material

Loam: The ideal soil made of sand, silt, and clay

Subsoil: The deeper layer of soil, usually lighter in color, which stores water

Topsoil: The top layer of soil, darker and more crumbly, where most nutrients exist

Good soil maintains a balance of water, air, organic materials, and nutrients through the natural cycle of growing plants. The roots take up minerals and water out of the soil, and uses those nutrients to produce fruit and leaves that then drop onto the ground to decay and return back into the soil.

Weed Indicators of Soil Type

Alkali soil	Coast—saltgrass (*Distichlis spicata*)
	Desert—chickweed (*Stellaria media*)
	Shepherd's purse (*Capsella bursa-pastor*)
	Blueweed (*Echium vulgare*)
	Gromwell/Puccoon (*Lithospermum*)
	Field peppergrass (*Lepidium campestre*)
	True chamomile (*Chamomilla matricaria*)
	Bellflower (*Campanula*)
	Salad burnet (*Poterium sanguisorba*)
	Scarlet pimpernel (*Anagallis arvensis*)
	Bladder campion (*Silene latifolia*)
Poor drainage	Dwarf St. John's Wort (*Hypericum*)
	Horsetail (*Equisetum*)
	Silverwood (*Potentila anserina*)
	Creeping buttercup (*Ranunculus repens*)
	Mosses
	Sumac (*Rhus integrifolia*)
	Curly dock (*Rumex crispus*)
	Sorrel (*Rumex acetosella*)
	Hedge nettle (*Stachys palustris*)
	Mayapple (*Podophyllum peltatum*)
	Thyme-leaved speedwell (*Veronica serpyllifolia*)
	American hellebore (*Veratrum viride*)
	White avens (*Geum album*)
Acid soil	Spurrey (*Spergula arvensis*)
	Corn marigold (*Chrysanthemum segetum*)

	Sow thistle (*Sonchus arvensis*)
	Scentless mayweed (*Matricaria inodora*)
	Plantain (*Plantago*)
	Lady's thumb (*Polygonum persicaria*)
	Goose tansy, rough cinquefoil (*Potentilla monspeliensis*)
	Wild strawberry (*Fragaria*)
	Rabbit-foot clover (*Trifolium arvense*)
	Horsetail (*Equisetum*)
	Dock (*Rumex*)
Slightly acid soil	English daisy (*Bellis perennis*)
	Sorrel (*Rumex*)
	Prostrate knotweed (*Polygonum aviculare*)
Very acid soil	Knapweed (*Centaurea nigra*)
	Hawkweed (*Hieracium*)
	Silvery cinquefoil (*Potentill argentea*)
	Horsetail (*Equisetum*), if swampy type
Heavy soil	Coltsfoot (*Tussilage farfara*)
	Creeping buttercup (*Ranunculus repens*)
	Dandelion (*Taraxacum officinale*)
	Plantain (*Plantago*)
	English daisy (*Bellis perennis*)
	Broadleaf dock (*Rumex obtusifolius*)
Light/sandy soil	Spurry (*Spergula arvensis*)
	Corn marigold (*Chrysanthemum segetum*)
	Sheep's sorrel (*Rumex*)
	Cornflower (*Centaurea cyanus*), especially when flowers are pink
	Small nettle (*Urtica urens*)
	Shepherd's purse (*Capsella bursa-pastoris*)
	White campion (*Lychnis alba*)
	Maltese thistle (*Centaurea melitensis*)
	St. Barnaby's thistle (*Centaurea solstitialis*)
Hard/crusted soil	All chamomiles
	Mustards
	Morning glory
	Quack grass (*Agropyron repens*)
	Goosefoot (*Chenopodiums*), no matter how bad the soil is
Salty soil	Russian thistle (*Salsola kali*)
	Sea aster (*Aster tripolium*)
	Asparagus (*Asparagus officinalis*)
	Beet (*Beta*)
	Shepherd's purse (*Capsella bursa-pastoris*)
	Mustards

Subsoil Indicators

If you dig down two to three feet deep and look at the subsoil, it will indicate the health of the topsoil. You can determine a lot by observing the color of the subsoil.

Red/yellow: Lots of iron oxides, good drainage, acidic soil, common in warm climates

Blue/blue gray: Lack of oxygen and poor drainage, common in thick layers of clay

White to ash gray: Nutrients and humus have leeched away, acidic and/or sandy, common under pine trees

Medium brown throughout: Good drainage

Pale or very similar to topsoil: Poorly developed soil, topsoil may have been removed

Dark brown: Abundant decomposed organic matter, usually where wetlands have been

Patches or streaks of colors: Pockets of poor drainage, different soils—plants may grow poorly from being waterlogged or lack of nutrients

---

The Best Possible Soil

Matter and organisms: 5 percent of soil—add soybean meal, garbage, cottonseed meal, plant residue, sludge, manure and compost

Minerals: 45 percent of soil—add lime, green sand, granite, dust, nitrates in organic matter

Water: 25 percent of soil—sprinkle or irrigate, but use the water table when possible

Air: 25 percent of soil—make the soil loose with organic material and cultivation

---

Roots end at same depth: Layer of compacted soil, cemented layer, and poor drainage.

Heavy Metals

One major worry with growing food in the city is toxins in the soil. If you are doing container planting with purchased soil this is usually not a problem, but in backyards and vacant lots it can be. When testing your soil, it's a good idea to test for heavy metals. Lead is a very common culprit, while mercury, arsenic, and cadmium are close seconds. You can get soil tests at your local university for a low price.

Before you go for the test, you should divide your growing area up into at least five sections (or more, depending on your budget), record these sections on a map, and give each one a number. Dig a few holes, each around a foot deep, within each section. Collect dirt from the side of each hole all the way to the bottom of the hole so that you get some from the entire depth using a plastic or stainless steel shovel (other metals will skew the results) and put that dirt into a bucket. Mix it all together, and then divide it into two samples of at least a cup of dirt each, and put each sample into a plastic bag, labeled with the number of the area. Make sure you clean your mixing bucket in-between each area. Send one bag from each area to the lab for testing and keep the other for backup.

The solutions to high heavy metal levels in your soil are pretty much the same as for repairing any other soil problem. Adding lots of compost will dilute lead levels, and including lime will make sure your pH balance is around 6.5, which will reduce the availability of lead to your plants. The danger of breathing in heavy metals when exposed to dirt particles in the air can be deadly, so using mulch is important. Make sure you wear gloves and wash your hands after gardening. Wash vegetables very well.

Cultivation versus No-Till

Cultivating (also called tilling) is the process of turning over soil with a plow or disc, or using hand methods such as a shovel to turn over the earth. There are many alternatives to cultivation, and these methods are outlined below. The goal in cultivation is to loosen up the dirt and remove any rocks so that plants will have the optimum environment. When cultivating with hand tools, people most often make a raised bed. A raised bed has the same amount of dirt, but it has been stirred and loosened so that it is fluffier, making it higher than the surrounding soil. Many raised beds are "double dug," a process of digging that makes it simpler to raise

the bed. More on how to prepare this is mentioned later in this section.

When cultivating, you would normally plant in rows based on the equipment you use. A tractor cultivator would allow you to plant single rows for a single row of plants, and a rototiller creates a standardized row that is usually around sixteen or thirty inches wide for city plots. More than one row of plants can be planted in it.

Urban farmers generally use raised beds at home in their permanent yard and rototillers on their leased or borrowed land. The reason is there's very little time and money needed in putting in a rototilled bed versus a double-dug bed. Generally, you just decide how much space, equipment, and time you have. With a lot of time and no space, do a raised bed because no-till is ideal. If you have little time and a big space, use rows and a rototiller.

Another option, besides double digging, is a boxed bed. This is a box made of wood, metal, stone, or other available material that is filled with soil. Landlords

▲ These no-till beds are built of rock and located in a suburban neighborhood.

like these if you have a rented house because they look nice. They also cost a lot more. If you are trying to decrease your initial startup cost, the wood frame does very little for your farm except to add aesthetic value. Even if you are putting in hoops to hold up plastic, they can be driven directly into the ground.

That being said, boxed beds are just a large, stationary form of container gardens, and they are awesome. Containers of any kind are better than boxed beds because they are completely mobile, which means you can take them with you when you leave. You can also move the plants around to better locations if you need to. Any kind of container with drainage can be used, which makes it possible to find something free or cheap. Old or new plastic planters, ceramic planters, old bathtubs, even large plastic and burlap sacks can work well.

Double Digging

Cover the plot six inches deep with your fertilizer or compost. Dig a trench as deep as you need, then dig another trench next to it. Put the second trench's dirt into the first trench, dig a third trench and put that dirt into the second trench, and so on throughout the plot. If there is a lot of grass in the dirt, shake apart the clumps before throwing them in the trench. Let the plot sit a few days, and then break up the clods with a hoe. Water the area and see if it is higher than the ground around it. If you did it right, it should be at least a few inches taller than the rest of the area. You would never want to walk or smash down the soil in the bed.

Corn, watermelon, pumpkin, and other similar types of plants can be planted in hills rather than beds. In the desert, your "hill" will actually be a hole that can catch water. In wet areas, make a hill six inches tall and plant your seeds in the top of it. This method uses a lot of space, but works especially well in clay-type soil.

A Brief Introduction to Natural Farming

Natural farming is a technique of growing plants without plowing, cultivation, or sometimes even fertilizer. After the initial transition from cultivation to natural (which can take a few years), the promoters of this method say that the plants are healthier and have fewer pests and less disease than cultivated plants.

The method is as follows:

1. In the fall season, put clover, vetch, or alfalfa seeds in a tray, pour fine powdered clay over them, and spray with a thin mist of water. Roll them around until they are well coated with clay. This makes pellets one inch in diameter.

2. Spread the pellets (also called seed balls) over the field. Cover the field with straw by throwing it down in all directions. This is your ground cover.

3. Let ten ducks per quarter acre forage loose in the field.

4. Plant your grain crop (with clay-covered seeds) when the preceding crop is ripening. After harvest, sow white clover with the grain as a ground cover. Cover with straw.

In an orchard, the soil would be treated the same as for fields, except you would cut weeds and tree sprouts with a scythe. For vegetables, the techniques differ according to climate. Wait for rain that will fall for several days in early spring. Try to plant so sprouts will grow before the weeds. Cut some of the weeds, throw down clay-covered seeds (seed balls), and lay the weeds on top as your ground cover instead of straw. You may need to cut the weeds back two or three times. Let chickens scratch through the garden—they won't eat the seed balls because they look and taste like dirt, but they'll eat most of the insect pests. Plant clover in late summer or fall to keep back weeds.

An alternative to total natural farming is the sandwich method. Before planting, layer barn litter, compost, grass clippings, chopped leaves, and wood ashes between layers of peat moss. Water it well, until it is damp and spongy. Then plant your seeds in it. You could also combine methods to use a cross between totally natural and a little bit of cultivation. Double dig a well-developed and fertilized bed to make a raised bed. After this, it must never be dug or walked on again. Put earthworms in the bed to do natural tilling, and then use the seed balls to plant and regular ground cover to control weeds.

The Art of Mulch

Mulch is any cover put over the soil to prevent weeds from growing and/or to insulate the ground (not all soil needs this). Put mulch on after the ground has thawed, unless you are growing strawberries. Strawberries grow best in cold, wet ground, so for them you can put it on before the ground has thawed. Use organic material, newspaper, or cardboard to cover all around your plants. Mulch keeps water in the soil, smothers weeds, and if you use organic material, fertilizes the ground. Some people swear by it because you may never have to weed.

There are many ways to mulch. The method described here is called *sheet mulching*, because you layer different sheets of materials together, something like a layer cake. Sheet mulching stops every kind of weed, saving you even more work.

1. Plant your largest trees and shrubs. If you get this out of the way, you won't have to go back and dig through your mulch later.
2. Cover the area with a sprinkling of dolomite, and if the soil is clay, add gypsum as well. Add any nitrogen that you can get, such as composted manure fertilizer or kitchen compost.
3. If you have some spare hay that is unfit for any other use, use it to make a layer one or two inches (2.5–5 cm) high. If the area was covered with tall weeds, you can simply cut those down and leave them lying on the ground instead. One warning about hay: do **not** use hay that has gone to seed. You will essentially be planting grass in your garden, and grass hay is tough to eradicate.
4. Cover the whole thing with a layer of cardboard, newspaper, old drywall, nonsynthetic carpet, felt underlay, or any material that is very hefty but will break down eventually. Don't allow even the smallest hole. If you do have to work around a tree, make sure the layer hugs the plant very tightly. This layer should be a half inch to one inch (1.5–2.5 cm) tall, with any noncompostable materials like staples or plastic tape removed.
5. Water it very well until it is completely soaked.
6. Add eight to twelve inches (20–30 cm) of old straw from a horse stable, old chicken coop sawdust, raked or old mashed-up leaves, seaweed, or seagrass. These all contain vital nutrients and can be moist. According to the composting principles (see the previous chapter) these should be brown materials rather than green,

which would turn into mush and smell bad.
7. Water everything again until it is well soaked.
8. Now you can add another one to two inches (2.5–5 cm) of compost and manure, and another two inches (5 cm) of dry material like straw or leaves. Sometimes sourcing all of this organic matter can be a challenge, but there's really no wrong way to do this, so don't sweat it. The only rule of thumb is the thicker the better. If you don't have enough material, you will have to make your bed smaller rather than try to spread it thin over a wide area.
9. Plant your largest seeds, potatoes, seedlings, and small potted plants. To do this, make a hole through the mulch to the sheet material. Use an old axe or knife to cut an X in the carpet or cardboard, or whatever you used as your sheet layer, and use your hands to put dirt in the hole. Stick the seeds, potatoes, or seedlings in the soil. If you want to plant tiny seeds, sprout them first and make a line rather than an X. You don't have to push the seeds down deep. Cover them with mulch, or for seedlings cover the base of the plant.
10. The roots won't do very well in the first year, but if you plant roots that grow down deep they will begin to break up the soil under the mulch. The next year you can grow more of them.
11. At the end of the first summer your soil will be immensely improved. You'll need to add a small amount of fresh mulch as the season continues, and in fact *annuals* (plants that are replanted every year) can tolerate food scraps direct from the kitchen layered under the mulch where worms will dispose of them immediately.

12. Keep the mulch area well watered. It will take frequent watering to keep the area moist but, as time goes on, it will get better at maintaining moisture.
13. As the seasons pass, the mulch will settle and shrink. Just keep adding layers and plant new things as the old ones are eaten, adding mulch layers each time.
14. If the weeds break through, just smash them under and add wet newspaper and a layer of sawdust. Grass will give up eventually. If a strong root takes hold, dig it up, fill the hole with fresh kitchen scraps and cover the hole with mulch. Make sure not to bury any fresh wood products. They need to be broken down in the air first. Mulch should be loose and light, with many different materials mixed together to help achieve that air texture.

Tilling

Rototilling is the only mechanical method of soil turning mentioned here because it is pretty much the only way to do this in an urban environment—unless you own a pair of oxen, which is unlikely. Rototilling should be done sparingly. It can be used to turn new ground (usually covered in grass), and to add inputs to the soil. Through trial and error most urban farmers using borrowed land end up using this as their income-producing system.

Repairing Very Bad Soil

If the soil is very compacted, you can loosen it with a tool, but don't *turn* it. Turning the soil means flipping it over, which you often see when someone plows up a grassy field. In a big area you can plow with an implement that just loosens up the soil, such as a *chisel plow*. A chisel

▲ This is a larger rototiller, which is more appropriate for urban farming than the small backyard gardener version.

plow cuts a slice into the ground without turning the soil over, allowing air and water to be absorbed. The first time you do this, plow four inches (10 cm) deep. The second time go down to seven inches (18 cm). On a slope, make sure to plow the channels diagonally instead of straight downhill, which will prevent water from flowing away

too easily. In smaller gardens use a garden fork rather than a plow.

If you planned out your land in the order outlined in this book, you have already controlled the flow of water by building swales. The pH test that you did on your soil indicates which kinds of plants to grow in that area, or what to add to change the pH.

Any soil, not just bad soil, should be improved by planting cover and green manure crops, or adding composted animal manure. You can also add *compost* (kitchen and yard waste that has decomposed) to the small gardens near the house. Soil that has been cleared most likely needs extra help because minerals have been leeching out.

Encourage worms and other beneficial creatures to live and grow in the soil. These are the best cultivators; they do their own composting and mulching.

Land that has experienced extreme erosion needs gentle treatment. Avoid grazing any animals on it for a long while, and it may be a good idea to plant a crop

Problem	Solution	Notes
Bare soil, no calcium	Calcium/silica	Cement dust, bamboo mulch, grain husks
Bare soil, no nitrogen	Nitrogen/potash	Leguminous tree mulch, manure
Desert soil, can't retain water	Bentonite	Volcanic clay absorbs water
Desert soil, too much clay	Gypsum	Allows water to penetrate
Salty soil	Raised beds	Salt leeches down
Potash deficiency	Comfrey, wood ash	Potash is potassium found in organic matter
No trace elements	Mulch/compost	High alkalinity prevents plants using minerals
Too alkaline (low pH)	Sulfur	Increase to 6.0–7.5 pH
Too acid (high pH)	Lime (calcium carbonate)	Decrease to 6.0–7.5 pH

The Ultimate Guide to Urban Farming

that has deep roots, such as daikon radish, chicory, and leguminous trees. Deep-rooted trees pull nutrients from the deepest layers of the soil, and their leaves can be used as mulch to return those nutrients back to the soil. The goal is to cover all exposed soil with quick-growing local species of trees and shrubs to prevent erosion.

Concrete, Gravel, and Other Beds

The only land available to you might be an empty parking lot covered in concrete, asphalt, or gravel. You may be tempted to rip up hard, man-made materials like this, but before you do, consider the benefits. The land will not grow weeds, it will be very warm as the concrete absorbs heat during the day, and it will be easy to move your wheelbarrow around. It's a lot better to build raised beds on top, but you will need access to soil and a truck. Some farms have done this by building beds on top of old pallets. They can move the pallets with a forklift if necessary, and they just put up wood sides with a waterproof sheet at the bottom and fill them with dirt. Alternatively, you could build *berm beds* directly on top of the ground. To do this, put down a layer of mulch at least a foot deep. This mulch can be wood chips, straw, cardboard, or other organic waste. Then add a layer of compost and manure about two to three feet high. Water everything down, then add more in the low spots. As you continue planting throughout the season, you may find that your beds will shrink and fall apart without a retaining wall. Between each planting you will need to use your tools to push the beds back into shape and probably add more soil/compost.

COMPOST

"You are not a beautiful and unique snowflake. You are the same decaying organic matter as everyone else, and we are all part of the same compost pile."
—*Chuck Palahniuk,* Fight Club

Compost is the ideal and optimum material for plants. It has all the microbes, bacteria, and nutrients that are needed to produce life, and it is absolutely necessary for success in growing vegetables. Compost is the future currency of the local food movement.

The Composting Process

Composting can be an art form, with the most skilled composter able to produce soil in just a few months. However, it also can be a simple process. If you put plant material in your compost pile, it will begin to decay, and in about six to eight months you will have usable compost no matter what you do. You just can't go wrong because bacteria, mold, and organisms are doing all the work.

There are three major compost methods:

Large Bin: Usually chosen by people who produce a lot of plant and vegetable waste from their big backyard garden. Three bins are built in a row, each one

around four to five square feet, and the new material is added only to the first bin. An average person makes two pounds of garbage a week, so this should hold enough for a four-person family. Put a brick or stone floor on the bottom. This pile is turned into the second bin after thirty days, if the pile is at least three feet tall and three feet wide. After another thirty days, the material in the second bin is transferred to the third bin. By the time the pile has spent thirty days in the third bin, it should be done. This method takes two to three months to make usable compost.

Black Bin: Also known as the backyard composter, this is a black plastic bin that has two features: a lid on the top and a removable door on the bottom for removing compost. There are many different styles of composters, but as long as they are made of black plastic and have some ventilation holes and doors, they all work the same way. These bins are appropriate for kitchen scraps and smaller quantities of garden waste. The bin should sit on bare soil so that organisms from the ground can move into it; it should be near enough to the back door that you can easily add scraps from the kitchen; and it should be in direct sunlight, as the heat from the sun helps to break down the plant material. It should not be up against the house or fence, because spillage from frequent

dumping of waste can damage nearby structures. Once full, the bin should be turned every couple of weeks for aeration. This method takes six to twelve months to make usable compost.

It's a good idea to add material to the bottom of any kind of compost pile or bin that can make an air space and help with aeration. Hedge trimmings, sticks, twigs, and other woody materials can work well for large bins and black bins.

The Carbon/Nitrogen Ratio

Everything has a carbon/nitrogen ratio, or *C:N ratio*. The ideal compost pile has a 25–30:1 C:N ratio, and each material has its own ratio. Table scraps are normally 15:1 and manure is 20:1. It can be a bit complex to keep the pile at this perfect ratio. The easiest way to keep track of this is to categorize everything into Greens and Browns. Greens are the nitrogen group and include fresh, green plants; fruit and vegetable scraps; seaweed; and fresh manure. Carbon is the Browns, including dry, dead material like paper, sawdust, and leaves. Simply add equal amounts of each to keep the right balance and allow the mixture to heat up. If you have the time, chopping everything up into tiny pieces speeds up the decomposition process.

For most people, food scraps and kitchen waste make up about 30 percent of their garbage. Keep two buckets under the kitchen sink, and add all of your food scraps *except* meat and dairy. Orange peels, banana peels, and eggshells should be crushed and cut up so that they will break down faster. You can also add small paper scraps. Once a bucket is full, dump it out in the compost heap. Once you've collected all of this material together, it will begin to heat up, usually within twenty-four hours. Organisms will gather to feast on the materials you brought for them, which will make the temperature rise. During this time, it is best to leave the pile alone and just keep adding more on the top. It should rise to over 120°F (49°C). When these microorganisms are done, the temperature will drop and you can stir it to get some oxygen mixed in.

You will know your compost is ready to harvest when it looks like soil. It should be dark and crumbly and have very few signs of food scraps. The material will have shrunk down quite a bit and won't smell bad anymore. You can add this material straight to your garden by stirring it into the first inch of soil or just sprinkling it on top.

Sometimes compost piles don't heat up, which is fine. It just takes longer. This is usually because the pile hasn't gotten bigger than three feet square, or it might be too wet or too dry. It should be about as moist as a damp sponge. Another problem that often arises is an imbalance of materials. This is not an exact science and you don't have to measure it, but experience will begin to tell you whether you need more Brown or Green materials. Often, if the pile is too wet it needs more Brown materials, and if it is too dry it needs more Green materials. It helps to chop up all the materials before adding them, so they'll break down faster.

Compost Troubleshooting

All garbage will compost, but troubleshooting your initial observations can make it happen quicker.

Pile is cool and dry: Add water until the center is evenly moist. In dry places, water and cover the compost.

Pile is cool and wet: Make sure pile is at least three square feet and four feet tall.

Large pile is cool and wet: Add alfalfa meal, manure, or fresh grass clippings, and stir.

Pile cools before composting: Turn and stir.

Pile smells bad and is wet: Add shredded newspaper or straw and stir. Cover to hide from rain.

Pile does not compost completely: Add newspaper or straw and stir.

Teas for the Soil

A liquefied tea is an efficient method of adding nutrients to the soil quickly. If you are adding mulch and composts to your soil anyway, this type of tea may not be necessary, but if plants fall victim to bad weather or attacks, this can help save a crop. It is particularly useful for tomatoes and peppers.

1. Mix one part manure to three parts water.
2. Leave it to ferment for at least two weeks in a container with a loosely fitting lid.
3. Add ten to fifteen parts more water. It should look like weak tea but smell disgusting.
4. As you use it, keep adding more manure and water so you can have a continuous supply.
5. Extras can be added to the tea, including comfrey (which adds potassium), seaweed, or kitchen compost.

Green Manure

Green manure is an easy way to fertilize and is more practical for a big field, but it can be done on a small plot also. The process is basic: you plant something, cover it with dirt, and wait for it to decay. Plant rye grass, buckwheat, barley, pearl millet, oats, or alfalfa in your field, then wait until it is mature to cut it down. Plow it under and wait until it is composted before planting seeds. It will look like your other soil when it has composted. The following are the most common types of green manure:

Legumes: Sow in the fall so flowers will attract beneficial insects.

Alfalfa (*Medicago sativa*): Sow in spring or summer.

Bell or fava bean (*Vicia faba*): Sow in fall or very early spring—they can tolerate 15°F.

Clovers: Berseem clover (*Trifolium alexandrinium*) is a summer or winter annual. Crimson clover (*T. incarnaturn*) is winter hardy and easily tilled. Dutch white clover (*T. repens*) is easily cultivated. Red clover (*T. pratense*) is a quick-growing biennial that can be planted from spring through fall. Subterranean clover (*T. subterraneum*) is a cool-season reseeding annual, best for sowing under taller crops. New Zealand white clover (*T. repens*) is a hardy, long-lived perennial and is heat resistant.

Peas: Field peas (*Pisum sativum*) and Austrian field peas (*Lathyrus hirsutus*) sow in fall or very early spring. Cowpeas or Southern peas (*Vigna sinensis*) grow as a summer annual.

Vetch: Common vetch (*Vicia sativa*) grows in most soils. Hairy vetch (*V. villosa*) is tolerant of very cold weather. Purple vetch (*V. atropurpurea*) is less cold-tolerant, making good winter-kill mulch.

Composting Manure

Pile manure up outside in a pile at least six feet deep and no more than six feet wide, and as long as you want. Cover it with dirt, and let it sit all winter. If you use it before then, it may kill your plants. If you keep your animals inside, gather that bedding to harvest their urine. The best animal beddings are chopped straw, corncobs, wood shavings, or sawdust because they will decompose well. Don't let the pile get too dry.

Bat guano; human, dog, and cat manure; sewage; fresh manure; or horse manure that hasn't been thoroughly composted (it contains weed seeds) can **not** be used. It can spread disease and destroy your soil.

Fertilizing

Green manure, manure, and compost in various mixtures make fertilizer. Mix together your compost and manure, then spread it on top of your green manure. Plow or turn the soil over to mix them into the dirt. If you are using no-till methods, you can safely rake the compost into the first few inches of soil. Always put the manure into the soil immediately after you uncover it from the dirt to save the gases. Even the gases released by the fertilizers are good for your soil.

WORMS OR VERMICOMPOSTING

"It was the day of the worms. That first almost-warm, after-the-rainy-night day in April, when you bolt from your house to find yourself in a world of worms. They were as numerous here in the East End as they had been in the West. The sidewalks, the streets. The very places where they didn't belong. Forlorn, marooned on concrete and asphalt, no place to burrow, April's orphans."

—*Jerry Spinelli*, Maniac Magee

The Ultimate Guide to Urban Farming

▲ **Red worms at work.**

Worms that have three lips, five hearts, a brain, and red wigglers are bisexual, although it takes two worms to make babies. They are useful for fishing, bird food, composting, and gardens. They tunnel and loosen the soil, reduce soil acidity, free plant nutrients into the soil, and make your topsoil deeper.

Getting Worms

You will need two pounds of worms per one pound of food waste you produce. There are many types of worms and they can all be used for different purposes.

Garden worm: The kind you find in your backyard or garden, they turn white in water. Don't use these for your vermicomposting bin.

African nightcrawler: About five inches long but can grow much bigger, they like warm weather and die in cold air or water. They reproduce every two years.

Native nightcrawler: Found in your yard and garden, they require lots of soil. These worms like temperatures under 50°F, and don't like to be disturbed. They are best for the garden but not the compost bin.

Red worms: They consume lots of garbage, reproduce every seven days, and have alternating red stripes. At one and a half to three inches long, they are used for fishing and are the best for the compost bin.

Sewer worms: Mainly found in manure piles.

Raising worms is no more difficult than composting. It is especially valuable to urban dwellers and can even be done in an apartment. The worms need a covered bin with holes in the sides and bottom. It's a good idea to line it with a mesh fabric to keep the little ones from crawling out. Set the bin on a drainage tray because you'll be watering it like a houseplant. You will also need bedding material. You can use nice black soil, shredded newspapers, or a mixture of both.

1. This bin will be able to hold about two thousand worms. For two pounds of worms, use a twenty-gallon bin to hold the worms and a slightly larger one that it can fit inside.

2. Drill holes in the sides of the twenty-gallon bin. Drill twenty or more ⅛-inch holes in the long sides. Drill maybe ten holes in the lid as well. You only need one lid. You only need a few holes in the bottom, which are slightly larger: ¼ inch.

3. On the bigger bin, tape or glue in aluminum cans that will act as a pedestal. You have to keep the smaller bin off the floor for drainage and air circulation. Put one of the small bins in the big bin on top of your pedestal.

4. This is actually the most difficult part. You now have to tear up lots and lots of little pieces of cardboard and newspaper, unless you have a big bag of leaves handy. Even then you may want to add cardboard and newspaper just to have a mixture. It takes a long time to fill up a bin this size. Make sure there's no tape or staples in there. You

then have to moisten the whole thing so it's as wet as a damp sponge. This bin took at least a gallon of water.

5. You only need two more things before the worms arrive: some food waste and a scoop of dirt. Worms need some grit in their system and adding a little dirt will provide that. Food waste can be a few days old, sort of broken down already. Just push aside a hole in the bedding, dump it in, and cover it up. The rules for worms are a little bit different than your regular compost. It's better to chop things up a bit for the worms, and you can't add very much bread or citrus. Bread heats up the bin to an uncomfortable level for the worms, and they just can't handle much citrus and peels. Other rules are the same as your compost: no meat, no dairy, no oily stuff. You'll need to add at least a half pound of waste per day, per pound of worms, and more if you want them to reproduce. One more thing . . . don't throw out the water dripping into the big bin. It is

super-scrumptious compost tea that needs to be put in the garden!

6. Now go out and get your worms. You need the kind that are sold for fishing, not the kind that are probably in your garden already. These are red wigglers, redworms, or branding worms. You will need about a half a pound (0.23 kg) of worms, or five hundred regular-sized worms per cubic foot (0.03 cu m).

7. The worms will eat their weight in kitchen scraps *per day*. Bury your leftovers in the bedding after each meal and follow the same rules as for your compost pile: no meat or dairy products. Garlic and potato peelings are also a bad idea. They do need a steady diet of crushed eggshells sprinkled on top.

8. They don't like cold temperatures. You can keep them outside in the summer, but if it gets chilly at night bring them in.

9. Whenever you take the lid off the bin, the worms will dive down to the bottom because they hate the light. Every couple of months open it up and put them in some bright light. After ten minutes, scrape off the top layer of materials—the valuable worm castings

that are super great for the soil. When you see worms, wait another ten minutes and do it again. Repeat until you've gotten it all, and fill it up with fresh bedding again.

PRODUCTION STRATEGIES

"We live in a world we did not create and cannot control."

—*Timothy C. Weiskel*

▲ Marigolds planted with tomatoes deter nematodes and other pests.

The Ultimate Guide to Urban Farming

Desert Gardens

The desert provides unique challenges and opportunities because of the heat and shortage of water. There are three strategies that will contribute to your success:

1. Create successions that are self-shaded.
2. As your staples, use edible tree species that thrive in dry climates.
3. Use boxes and trellises as close to the house as possible or attached to it.

Your *staple* foods are olives, palms, citrus, avocado, apricot, banana, and maybe papaya. Staples are foods that provide you with half of your diet when they are in season because they are nutritious and produce lots of food. A gray-water marsh and other rainwater runoffs feed into swales directly to the trees, which also simultaneously shelter the house and create a cooler climate. Well-mulched raised beds made of mud, with the shade of wood trellises, vines, or trees, are planted with peas and fava beans in succession with celery, onion, carrots, spinach, tomatoes, peppers, and melons. Climbing vines like grapes can be grown year-round in succession as well.

Succession and Interplanting

Many people plant a garden, but once the seed or seedling is in the soil, they don't know what to do other than give it water and hope for the best. However, there's still more work to do. The next step for vegetables is to get the plant to flower, because without flowers there aren't any vegetables (unless it's a root crop). The next step for herbs and lettuces is to prevent flowering, because if they do, they will stop producing leaves.

As the season progresses, you will also need to replant. Eventually, no matter what you do, the lettuce will *bolt* and stop making more lettuce leaves for you. With careful planning, you can extend your growing season far beyond what it otherwise would have been.

In the following sample schedule, plants are grown in quick succession to maximize food production:

Before First Frost: Around two weeks before the first frost of the season, you can start sprouting some quick-growing, cold-hardy brassicas inside. These include cabbage and broccoli.

Last Frost Date: Once you know that the last frost has passed, plant cold-hardy plants outside. These include lettuce, kale, dill, radish, parsnip, mustard, arugula, and carrots. Plant these close together, especially the lettuce.

One Month After Last Frost: The radishes should be ready to eat, and the brassica seedlings can be transplanted into the space where the radish was. The other herbs and leaf vegetables should begin to be ready as well.

Early Summer: When the soil has warmed up, remove a few whole heads of lettuce and plant bush beans in the spaces where they were. The cabbages will begin ripening and the remaining greens will try to bolt, so make sure you harvest those leaves quickly to prevent this.

Fall: You can finally harvest parsnips, and as the plants begin to end you can put in fava beans or push garlic cloves into the ground.

You will notice that there is a great deal of crowding going on. Rather than waiting for lettuce to get big, you are removing the whole lettuce as it begins to crowd everything around it. Don't remove all the lettuce heads, but just enough to give other plants room. Then continue to harvest the leaves from the other plants.

Using Plant Communities

There is an alternative to simple succession and interplanting. Some farmers believe that plant communities are much more productive than the standard method because they seem to be more resilient and require less work. To do this, plants are arranged into communities, sometimes called *guilds*. These communities are a way of organizing plants around a central element based on *companion planting* (plants that grow best together) and their growing tendencies. This reduces competition with the roots of other plants, provides shelter from the elements, adds nutrients to the surrounding soil, and deters pests.

Companion planting has been used for thousands of years, with the most well-known being the Three Sisters—corn, beans, and squash—which were commonly planted by the native people of the Americas. The corn provides a support for the beans, and the squash shades the ground, preventing weeds from growing. Together they produce much more food per square foot than they could when spread out on their own. Another example is the apple tree community. Several rings of plants are grown under the apple tree's canopy. At the base of the tree is a legume like fava beans, followed by a ground cover of clover with dandelions and chicory mixed in, surrounded by a ring of comfrey, artichoke, yarrow, nasturtiums, and dill, enclosed by daffodils that stop the grass from encroaching. This community is more complex than the Three Sisters, but each plant serves a purpose: attracting bees,

providing mulch, fixing nitrogen, and preventing grass cover.

At times, some of these companion plants will need to be grown in succession to take full benefit of each other. For example, the corn must be planted before the beans and squash or the beans will grow too quickly and knock the corn over. Pay attention to the growing times and use your common sense when planning your garden. Observation is key in this process; keep track of what plants grow well together, and even what plants have sprung up on their own in close proximity to others. Even more apparent may be the plants that *don't* do well together. The success of a plant community often has less to do with some unseen force, and is likely to come from a directly observational characteristic (like the ones you wrote down in the beginning when arranging your elements). This includes attracting or repelling deer or birds, attracting or repelling different insect species, or the function of the leaves and whether they shade the sun or mulch the ground. The one invisible trait that you should also research is the roots. The plant may fix nitrogen, loosen the soil, or provide a buffer that protects other plants from natural herbicidal plants.

Plants can be placed strategically to:
- Attract predators—these plants provide food or shelter to friendly insects that eat pests;
- Sacrifice themselves—these plants attract pests so the other plants can be left alone;
- Trap pests—these plants attract and kill pests, or trap them so you can do it. Care must be taken to avoid providing a hotel for pests. Some

plants provide a nice home for pests to live all winter, giving them even more opportunity to destroy your crops in the summer.

- Provide nutrients—these plants can be grown to fix nitrogen or create other nutrients and friendly bacteria. They can also be cut down and left as a mulch below trees or between crops;
- Create shelter—these plants prevent frost, stop the wind, make mulch, and create microclimates.

Animals can also work together to provide strategic benefits:

- Foraging—when fruit falls on the ground, flies and other pests gather. Pigs and birds can clean these up, simultaneously fertilizing as they go;
- Insect control—birds that eat larvae and eggs from the bark of trees can be attracted with flowers and herbs. The flowers and herbs attract insects, which in turn attract birds;
- Slug and snail control—ducks can be allowed into the garden from fall to spring, where they will totally control the slug and snail population. They stay in the marsh in the summer;
- Predator control—foxes, deer, and rabbits can all be controlled with a couple of dogs. If the dogs are raised with chickens, they will leave the poultry alone and protect them from predators.

It can be difficult to plan communities of plants and animals when you are working with so much diversity. How do you make sure everything in a group works together without losing your mind? This is done through a *coaction study*. Coaction means "acting together," and so when you do a coaction study, you are finding out which plants act together for the most benefit. It's actually quite difficult to go wrong in this process, since many living things will grow together without any effect on each other at all. Your goal is to try to find the ones that help each other. This is important when you are working with many communities that all interact with each other in one mega-community.

1. When Species A and Species B are put together, they can affect each other in one of three ways: positively, negatively, or not at all.

2. Obviously, if both species were negatively affected, you wouldn't want to put them together in the future. Things are not so clear if one is positively affected and the other is negatively affected, or not at all. If Species A is benefited but Species B is negatively affected, Species A is a parasite.

3. Begin observing each species in a community to see how it is doing in relation to the species around it. For example, the age-old knowledge of companion planting tells us that these plants will affect cucumbers:

   − = *negative*

   += *positive*

   0 = *none*

4. We could also observe our local community to find out what plants do well together in our area. If we wanted to see if the companion planting rules hold true, we could observe gardens to find out how the cucumbers are doing and tally the results. The number of cucumber plants that do well can be written rather than a + or −. Here are some examples:

	Tomato	Carrot	Lettuce	Radish
Cucumber	–	+	0	+

	Tomato	Carrot	Lettuce	Radish
+	5	6	7	8
–	5	3		1
0		3	3	1

5. When the results are compared, they can be slightly different than expected. Cucumbers aren't supposed to do well with tomatoes and yet the results are fifty-fifty positive and negative. Lettuce is supposed to be neutral, and yet cucumbers seem to be growing well with them.

6. Once we have studied and researched these species and how they do within a group, we can begin to put them together so that each one has a +. Every plant should be benefited by something and be removed from any species that affects it negatively.

Companion Planting Table

Type of Plant	Does Well with	Does Poorly with
Amaranth	Sweet corn	
Anise	Coriander (improves growth and flavor) Cabbage family	
Apple	Chives Clover	
Asparagus	Parsley and basil (deters asparagus beetle) Tomato Nasturtium	Onion Garlic Gladiolus
Basil	Tomato Everything (improves flavor and growth) Asparagus Nasturtium Pepper	Rue
Bean	Carrot Cabbage family Beets Cucumber Corn Grain Pea (improves growth) Spinach Eggplant	Pole bean Wormwood Marigold

Type of Plant	Does Well with	Does Poorly with
	Mustard Potato Rosemary	
Beet	Bush bean Cabbage Lettuce Onion Kohlrabi Lima bean Radish Celery	
Borage	Tomato Squash Strawberry	
Broccoli Cauliflower	Sage	
Bush bean	Beet Carrot Cucumber Marigold Potato Cabbage Celery Corn Eggplant Lettuce Pea Radish Strawberry	Fennel Garlic Onion
Cabbage family	Sage (deters pests and improves growth) Celery Beets Onion family Chamomile Spinach Chard Pea (improves growth) Tansy (deters cutworm and cabbage worm) Anise Bean	Dill Pole beans Strawberry Tomato

Type of Plant	Does Well with	Does Poorly with
	Bush bean Cucumber Hyssop Mint Cabbage Rosemary Thyme	
Carrot	Leaf lettuce Parsley Tomato Sage (deters rust or carrot flies and improves growth) Chervil (deters Japanese beetle) Pea (improves growth) Radish (deters cucumber beetle, rust flies, and disease) Bean Bush bean Chive Flax Onion Pea Rosemary	Dill
Celery	Cabbage Bush bean Onion Spinach Tomato	
Chard	Cabbage	
Chervil	Radish Lettuce Carrot Grape Rose Tomato	
Chive	Carrot Tomato Apple Grapes Roses	

Type of Plant	Does Well with	Does Poorly with
Clover	Apple	
Collard		Tansy
Coriander	Anise	
Corn	Snap or soybean (improves corn growth) Cucumber Pea (improves growth) Potato Pumpkin Squash Amaranth Radish Sunflower Bush bean Pole bean Onion	
Cucumber	Bean Cabbage family Corn Pea (improves growth) Radish (deters cucumber beetle, rust flies, and disease) Bush bean Dill Lettuce	Sage Potato
Dill	Lettuce Onion Cucumber	Carrot Tomato Cabbage
Eggplant	Bean Four-o'clock Bush bean Pole bean Spinach	Potato Peppers
Flax	Carrots Potato	
Garlic	Roses Everything (deters aphids and beetles)	Asparagus Bush bean Pea Pole bean

Type of Plant	Does Well with	Does Poorly with
Grape	Chervil (deters Japanese beetles) Chive Hyssop Nasturtium Radish Tansy	
Hyssop	Cabbage Grapes	Radish
Kohlrabi	Beet	
Lettuce	Carrot Cucumber Onion Radish Beet Chervil Dill Pea Bush bean Pole bean Strawberry	Chrysanthemum
Mint	Cabbage Strawberry	
Mustard	Beans	
Onion	Beet Cabbage family Lettuce Tomato Carrot (together deters rust flies and nematodes) Dill Celery Cucumber Pepper Squash Strawberry	Asparagus Bush bean Pea Pole bean
Parsley	Tomato Asparagus Rose Carrot	

Type of Plant	Does Well with	Does Poorly with
Pea	Bean Carrot Corn Cucumber Potato Radish Turnip Lettuce Spinach Cabbage Bush bean Pole bean Pea Squash	Garlic Onion Wormwood Potato
Pepper	Four-o'clock Onion	Potato Eggplant Tomato
Pole bean	Marigold Radish Carrot Corn Cucumber Eggplant Lettuce Peas Radish	Garlic Onion Wormwood Bean Cabbage Beet
Potato	Bean Cabbage Corn Marigold Pea (improves growth) Horseradish or tansy (deters Colorado potato beetle) Bush bean Flax	Sunflower Peppers Cucumber Eggplant Pumpkin Tomato
Pumpkin	Corn	Potato
Radish	Beet Carrot Spinach Squash	Hyssop

Type of Plant	Does Well with	Does Poorly with
	Corn Pea (improves growth) Chervil Cucumber Lettuce Pole bean Grape Bush bean	
Rose	Chervil (deters Japanese beetles) Chive Garlic Parsley	
Rosemary	Cabbage Bean Carrots Sage	
Rutabaga Turnip	Pea (improves growth)	
Sage	Broccoli Cauliflower Rosemary Cabbage Carrot Strawberry Tomato Beet	Cucumber Rue
Savory	Beans Onions Cabbage	
Spinach	Pea (improves growth) Bean or tomato (improves growth with shade) Cabbage Radish Celery Eggplant Strawberry	
Squash	Nasturtium Radish (deters cucumber beetle, rust flies, and disease)	

Type of Plant	Does Well with	Does Poorly with
	Tansy Borage Corn Onion	
Strawberry	Borage or sage (enhances flavor, deter rust flies and disease) Mint (deters aphids and ants) Bush bean Lettuce Onion Spinach	Cabbage
Sunflower	Corn	Potato
Tansy	Cabbage Potato Squash Grape	Collards
Thyme	Cabbage family Tomato (with cabbage deters flea beetles, cabbage maggot, cabbage butterflies, Colorado potato beetle, and cabbage worms) Borage	
Tomato	Asparagus Basil Garlic Marigold Parsley Chervil (deters Japanese beetles) Spinach Carrot Chive Sage Thyme Celery Onion	Cabbage family Fennel Potato Pepper Dill Corn

Weeding

Some people don't weed at all, and others weed fanatically. Most people weed sometimes and do a medium job. The fewer weeds the better because your plants won't have to compete with them for water, sunlight, and nourishment. After a garden has been in the same spot a while you will have fewer weeds to pull, and this is where natural farming starts to really work. But no matter what, in the first few years you will have to weed at least a little bit.

Weeding is hard work and it's not fun, although some people find it therapeutic. Weed when the soil is damp, such as in the morning after the dew, or after rain. During hot, dry days the weeds hold on very tight to the soil. It is uncomfortable to weed bending over, so the most comfortable way is on all fours, squatting, or by sitting on something such as newspaper or kneeling pads available from the garden supply store.

INDOOR GROWING

"Gardens are not made by singing 'Oh, how beautiful,' and sitting in the shade."

—*Rudyard Kipling*

If you live in a place where there is no natural light and you don't have a backyard or greenhouse, you might need to grow plants inside. Some people do this hydroponically, where they use factory-made fertilizer in water and grow under lights. In a situation where you couldn't get that special fertilizer, your plants would die quickly, so the best method is really to use soil and electric light.

Basically, you need to find soil and make your own compost using your food scraps and, perhaps, a composting toilet. Your plants need to be put in containers: the cheapest way to get them is from thrift stores or to use materials you have. Square containers save space and hold more soil. You will also need water trays for under the plants, tables for the containers, and electric lights. Plants need certain colors of light in order to optimize their growth. Red and blue are the most important, and these must be balanced, or plants will get too short or too long and skinny.

Types of Lightbulbs

Incandescent bulbs: The normal lightbulb that you use in your house. These work well but do not have enough blue light. They can also be somewhat hot, so if you put the plant too close it can scorch.

Fluorescent bulbs: Produce three to four times as much light as incandescent with the same amount of energy and are relatively inexpensive. They are sold in cool white (with more blue and yellow-green colors) or warm white (with more orange and red colors), so the best way is to put the two together in a fixture so the plants get all the necessary colors. Full-spectrum fluorescent bulbs contain all the colors and

are the best kind for a small farmer. They last twice as long.

High intensity discharge lamps (HID): Used by professionals, they use a lot of energy and are more expensive, and some produce so much light that protection must be worn around them.

Garden plants and vegetables require fourteen to eighteen hours of light each day. Then the lights must be shut off so that the plants can rest. If you notice the plants seem to be reaching for the light or have huge leaves, they may need more light.

Rotate the plants each week to get an even distribution of light, and use white trays or foil reflectors. Clean fluorescent bulbs once a month to clear them of dust. Feel your plants—if they are warm, the light is too close. And just like any other garden, you must water the plants and fertilize the soil with your compost.

Strategies

There is no magic formula to indoor growing. Some crops, like microgreens, need a lot of ventilation because of the high levels of moisture needed to grow them. A couple of well-placed fans may be all that is necessary. Mushrooms are another easy indoor crop with climate requirements. They need damp, dark, cool spaces such as a basement or closet. Each mushroom requires a different growing medium, which is specified when you purchase the spawn.

These are common profitable indoor crops:

- Microgreens
- Shitake, oyster, and button mushrooms
- Lavender
- Hardneck garlic
- Herbs
- Heirloom tomatoes
- Veggie starts

▲ My own microgreen system used Costco shelving and fluorescent fixtures on a timer.

Some indoor farms are aquaponic operations inside huge warehouses. They use fish tanks to raise fish that produce a rich fertilizer that is cycled directly through plant roots, eliminating the need for soil. Leafy greens do particularly well in this kind of large-scale indoor scenario under lights or in a greenhouse. For an apartment dweller with limited space, mushrooms and herbs are your best bet, as they are low energy and don't require the light that the other crops do.

PLANTING

"It always amazes me to look at the little, wrinkled brown seeds and think of the rainbows in 'em," said Captain Jim. **"When I ponder on them seeds I don't find it nowise hard to believe that we've got souls that'll live in other worlds. You couldn't hardly**

believe there was life in them tiny things, some no bigger than grains of dust, let alone colour and scent, if you hadn't seen the miracle, could you?"

—L. M. Montgomery, Anne's House of Dreams

Where to Buy Seeds

The last place you will want to buy seeds is from the local grocery store. If you are dedicated to growing organically, even if you can't certify, you are obligated to purchase organic seeds. Finding a reputable source of bulk organic seeds can be a major task. Seeds are often your most expensive input and must be prepurchased before the growing season. If you are succession planting, you have to plan ahead and buy enough for multiple plantings. Ordering midseason doesn't work because seed suppliers often run out

of the variety you might be looking for, so order early and well.

For a two-acre plot of land using succession planting, my own seeds cost somewhere around $1,700 for more than seven hundred thousand organic seeds for one season, which included a large quantity of microgreen seeds. On page 101 is an example of a seed record that would allow you to keep track of your current seed inventory. You won't run out of seeds if you keep track like this and make orders routinely.

Reading the Seed Packet

Seed packets are like real estate listings—brief, and with vague terminology that is intended to help you make the right choice. The following phrases are found on seed packets and indicate climate needs:

Frost-tender: Can't survive even a light frost. Don't plant early, and make sure the soil temperature is warm enough.

Semihardy: Can live through a light frost. Won't survive a heavy frost, but you can plant right around the last frost date for your area.

Hardy: Will survive early spring and late fall frosts, and can be planted a couple weeks before the last frost date.

Latin Names

Plants are identified with both a popular name and a Latin name; for example: cabbage is also called *Brassica oleracea.* Latin names are the scientific way of identifying each species, and are much more accurate than popular names. When buying seeds, always make sure that the Latin name is specified so you know exactly what you are getting.

All living things are classified into five kingdoms. Plants are part of Kingdom

Type	Quantity	Date	Supplier	Price	Variety	Used
Leeks	7,350	Jan. 15	West Coast	$49.95	Tadorna Organic	7,000
Brussels Sprouts	1,000	Jan. 15	West Coast	$37.95	United	900
Bush Beans	1,000	Jan. 20	William Dam	$17.25	Slenderette Organic	500
Peas	10,400	Jan. 20	William Dam	$99.50	Norii Organic	4,300
Beets	6,000	Jan. 20	William Dam	$46.50	Touchstone Gold Organic	6,200
Carrots	12,000	Jan. 23	Veseys	$50.90	Napoli Organic	12,000
Cucumber	2,000	Jan. 23	Veseys	$24.95	Marketmore 76 Organic	700
Bunching Onion	15,000	Jan. 23	Veseys	$36.95	Parade Organic	13,300

Plantae, fungus is in Kingdom Fungi, and animals are part of Kingdom Animalia. There are also Kingdom Monera, made up of microscopic single-celled creatures such as blue-green algae, and Kingdom Protista, which are similar to Monera but have complex cells, such as seaweed.

To further categorize things, each kingdom has several phyla. The mushroom group is phylum Basidiomycota, the yeast group is called phylum Ascomycota. In the Plantae species there are two groups. One is for conifer, moss, and fern-type plants called gymnosperms. The other is the group of plants that animals (including us) eat, called angiosperms.

Angiosperms come in two classes: Monocotyledoneae and Dicotyledoneae, or monocots and dicots. Monocots have seeds with one leaf, the plant has narrow leaves, and the flower has parts in multiples of three. Dicots have seeds with two leaves, the plant has broad leaves, and the flower has parts in multiples of four or five with large colorful petals. Within each monocot and dicot family there are genera. The genus of a plant identifies exactly which species it is.

The Latin name of the common foxglove is *Digitalis purpurea*. In Latin, the genus is most often capitalized, and the species name is not. Books and seed catalogs use abbreviations, so it could also be listed as *D. purpurea*. It is important to get the right variety, since this wild variety is used to make a drug that treats heart disease. Other foxgloves (*D. mertonensis*, *D. grandiflora*, etc.) do not have the same properties. Here is an example of all the classifications of foxglove:

Kingdom: Plantae
Phylum: Angiospermophyta
Class: Dicotyledoneae

Family: Scrophulariaceae (popular name: figwort, related to foxgloves)
Genus: *Digitalis*
Species: *purpurea*

Monocot families you should know

Arecaceae: palm family—coconuts and palms (equivalent of wheat in a tropical area)
Graminae: grass family—wheat, bamboo, corn, rice
Liliaceae: lily family—onion, lily, tulips
Musaceae: banana family

Dicot families you should know

Apiaceae: carrot family—carrot, parsley
Asteraceae: sunflower family—dandelion, sunflower
Brassicaceae: cabbage family—cabbage, cauliflower, kale, turnip
Cucurbitaceae: melon family—cucumber, melons, squash, pumpkin
Fabaceae: pea family—peas, peanuts
Lamiaceae: mint family—lavender, mint
Leguminoseae: legume family—alfalfa, bean, peanut, pea, soybean
Poaceae: grass family—wheat, barley
Rosaceae: rose family—rose, apple
Solanaceae: nightshade family—pepper, potato, tomato

Planting Seeds Outside

Seeds germinate when the soil is within a certain temperature range. Usually this is 50–70°F (15–20°C). So, when we are waiting for spring planting, what we are really waiting for is the soil to warm up to 50°F. We can warm the soil with plastic to speed up the process. Some seeds, like apples, wild rice, and some berries, require cold temperatures to sprout. They should be stored in the refrigerator all winter, with the wild rice also kept in water. When they begin to sprout they can be planted outside.

Some seeds also need light to germinate. Carrots, lettuce, spinach, parsley, parsnips, and beets can be thrown directly on the ground rather than pushed into the soil, but they are likely to be eaten by birds. Instead, you can soak them overnight and let them sit in the light before putting them into the ground. Larger seeds are able to germinate in the dark, and some (like parsley) need the dark.

Depth is another factor that gardeners fuss over. It is usually recommended that you bury the seed four times the largest diameter of the seed. The deeper you push the seed down, the darker and wetter it will be. A shallow seed depth is likely to have more light but also dry out. The seed packets will have a recommended planting depth that should be followed, and sprouting seeds indoors will be more successful than sowing directly in the ground.

Propagation

Propagation is the process by which plants reproduce. Some plants, like potatoes, bananas, and many kinds of herbs, don't use seeds. They may grow just from their own roots, or even by planting a piece of the fruit.

Plants with many stems, or a root clump (a large mass of smaller roots), can be propagated this way. Take a plant that is at least two years old, dig it up, and soak the roots overnight in water. This will soften them so you can gently pull the plant apart without tearing it. Plant pieces separately.

Lemon balm, comfrey, mint, horseradish, and others can be propagated by taking a plant at least two years old, and cutting off a piece of root at least two inches long. These plants have a single, fat, almost woody root, and you need to find a big, healthy-looking one. Plant it in the ground like you would a seed.

Layering is another way of propagating. For some shrubs, just bend a branch over and cover the middle of it with dirt (leaving the end out). The middle will grow roots into the ground and you can cut the plant away from the old one.

Tubers have their own special way. Get a nice potato, hopefully one that represents your ideal potato, and cut it into pieces. Each piece should have an eye. On a potato the eyes are darker brown spots, which indent into the tuber making it lumpy; on other tubers they are somewhat similar. When you cut be careful not to damage the eye. Plant each piece and a potato plant will grow from it.

All the other plants you will be dealing with must be started from seeds. However, to intensify your planting schedule, it's a good idea to start some in trays inside, ideally in a greenhouse. A "flat" is a plastic planting tray with a standard size of ten by twenty inches. They are also called 1020 trays, and come in a variety of styles. You usually need two kinds, a plain tray that catches water, and a "plug" tray. The plug trays vary in size so that your 1020 flat can hold different numbers of plants. You can just plant in the open flat, but it's a lot easier to remove the fragile seedlings from a proper plug tray.

When dealing with trays, and this is true for microgreens as well, you should sterilize the trays between each use. This is done by soaking in a bleach solution of one-half teaspoon of chlorine bleach per five gallons of water. Some people have recommended spraying the solution onto the trays, or using a hydrogen peroxide solution, but in my experience those methods do not adequately prevent mold from growing.

Each plant in the plant guide has its own propagation directions, but here are some rules of thumb:

1. A heavy feeder such as tomatoes needs more space, so larger plugs are a good idea. The most common is called a six-pack, which has six plugs and fits six to a flat, giving you thirty-six plants.

2. Most seeds need constant moisture for optimum germination. Beans, cilantro, corn, cucumber, melons, okra, peas, and squash can be soaked overnight and planted the next day.

3. The easiest way to plant the seeds is to place them on top of damp soil, then push them down with your finger. If you water after the seeds are planted in dry soil, they often rise to the surface and can be lost.

4. Big seeds can be planted one to a plug, but small seeds need a few in each spot and then can be thinned later.

5. Make sure you label your seedlings! This can be done with cheap plastic stakes or Popsicle sticks that you can write on with a marker.

6. Water your seedlings when the soil just begins to dry out. This means that your plants should be spongy damp, not soaking wet and not bone dry. The best strategy is to soak the soil once, then make another pass to make sure you got it enough.

Your seedlings are ready to plant out when they have one set of "true leaves," strong roots, and the soil is the right temperature. When a seed grows, it has a set of baby leaves that die off when the first set of true leaves form. The roots will also branch out toward the bottom of the pot. When this happens, it's time to "harden off" the seedlings by putting them outside longer and longer each day so they get used to temperature and weather changes. Bring them outside to a sheltered place like a porch during the day for longer and longer periods of time, a couple of hours the first day, a few the next, and by the end of the week they can be out all day and night. Remove them from the seedling trays without disturbing the roots and pop them into the ground so that the soil comes up almost to the first leaves. This helps them establish their root base faster.

Replanting Annuals the Easy Way

These techniques only really work well in a mild climate, but you could use cold frames or the greenhouse to try to do this as well. Certain plants can be made perennial:

Leek: Allow some leeks to go to seed. At the end of the season you can dig them up; you will find small secondary bulbs growing off the base of the stem. These can be planted just like onions. Alternatively, at harvest, you can cut off the leeks at the ground instead of pulling them out, and they will grow a second time.

Garlic: Garlic can be a perennial. If you leave it in the ground for a couple of years, it will begin giving you an everlasting crop of garlic.

Broad bean: Large pods often grow near the base of a broad bean plant, where they can be left on the ground to dry. In late summer, mulch over the top of them with straw and they will sprout in the fall.

Potato: In the fall, leave seed potatoes in the ground and mulch them well. They will sprout in the spring.

Lettuce: Allow lettuce to *go to seed* (don't harvest it, just let it flower) and it will scatter seed and replant itself.

Fruit and melon: Tomatoes, pumpkins, and other melons can be left in the garden and covered with mulch at harvest time, where they will rot and spill their seeds. These will grow and effectively replant themselves.

Carrot: When you eat carrots, keep the tops and store them in a dark cool place. They will begin to sprout and you can set them out to grow.

Cabbage: Cut the stalk high on the plant, leaving a few leaves. Little cabbage heads will spring up out of the stalk, which can be eaten or replanted.

TOOLS

"Your first job is to prepare the soil. The best tool for this is your neighbor's motorized garden tiller. If your neighbor does not own a garden tiller, suggest that he buy one."

—Dave Barry

Growing plants requires a number of specialized tools other than the ones I've already discussed, such as, a rototiller, fridge and freezer, and canning supplies. A few tools that are absolutely necessary to have on hand are:

- Shovels, both flat and pointed
- Hoes
- Bow rakes
- Pruning shears
- Trowels
- Scissors
- Large plastic bins or tubs
- Spray nozzles
- Tape measure

These tools should be high quality because they will see a lot of use. Shovels, rakes, and hoes often break where they are attached to the handle. The bow rake is used for shaping raised beds. The tines are for moving the soil and the flat side is for smoothing the top of the bed. You will need the scissors and pruning shears for harvesting because scissors work great for herbs and baby greens. The tubs are for harvesting and washing produce. Spray nozzles for hoses also tend to wear out very quickly when irrigating twice a day. Get a high-quality one with several spray settings.

These are some nice things to have:

- **Dibber or dibbler:** This is a device for making holes for seeds. It can be as simple as using a chopstick that is marked at different depths to a large

board fitted with pegs evenly spaced from each other. The board is set into the bed and the pegs "dibble" the holes.

- **Harvest knives:** A sharp, curved blade knife with a small handle is most recommended. It is used to cut squash stems or go quickly through a lettuce bed for full-grown lettuce.
- **Flats:** A "flat" is a black plastic tray that is used to grow small plants and seedlings. These are often cheap and not very durable, but can be used for microgreens, sprouts, seedlings, and even harvesting delicate plants.
- **Machete:** When you have a lot of garden waste, it needs to be chopped down for compost. For example, snow pea vines can be pulled at the end of the season, chopped up, and thrown in the compost to be broken down much more quickly.
- **Wheelbarrow:** This is pretty much a necessity, but they are expensive unless you can find one used. A small four-cubic-foot wheelbarrow is perfect for weaving in and out of tight spaces, and you need it to dump dirt, fertilizer, or compost into a bed.
- **Boots and waterproof clothing:** A pair of insulated waterproof boots is a great investment. If you take care of your feet, you will help prevent fatigue and injury, and be a little less tired at the end of the day. High-quality weatherproof clothing goes a long way.
- **Restaurant salad spinner:** A large salad spinner is a necessity if you are selling prewashed baby greens by the bag. You can do this with small, cheap salad spinners, but believe me, the time wasted is immense. Some people have converted an old washing machine into one, which is always an option.
- **Flame weeder:** A fuel-powered flame thrower (essentially) for killing weeds very quickly. This may seem like overkill (no pun intended), but when your time is precious, this is a great no-till weeding method.
- **Seeder:** This is a device that was invented a long time ago and has been brought back into production for the small farmer. It has disks that allow seeds to fall through at specific intervals. You can walk upright down the rows holding onto the handles to steer while it rolls along, dropping the seeds at just the right distance.

The city is not always the best place to find used garden supplies, but you might be surprised. Sometimes it can be worth it to visit rural agricultural auctions, farm-supply stores, and farm garage sales to see what you can find. Your garden center in the city will be more expensive, but if you are going to invest in new tools, it is totally worth it to buy a more expensive, high-quality item. Japanese and English tools are usually the best.

Hardware stores are another good option. Some of the best deals on tools and plastic that I have purchased were from locally owned hardware stores. Sometimes they'd give me a deal once they found out what I was doing with the tools. I have been offered plastic and other tools at wholesale price as long as I purchased the lumber (also a good deal) for my beds from them.

It's important when you are purchasing supplies to always shop around, and the Internet makes it possible to get the best price. While my plastic sheets might come from a local person, my deer netting and drip hose could be purchased on a spindle from a wholesale supplier on the Internet. Packaging supplies, organic pest control, worms, ladybugs, giant salad spinners, and other tools are all available online.

INCREASING PRODUCTION

"Despite eating more than ever before, our culture may be the only one in human history to value food so little."
—*Alisa Smith,* Plenty: One Man, One Woman, and a Raucous Year of Eating Locally

Becoming Totally Self-Sufficient

Although this book focuses on commercial production, many urban farmers are concerned with their own diet.

Planning is extremely important when you plan to live completely off your garden. You will need to be familiar with the food pyramid, and be able to grow a variety of things in order to maintain your family's health. With a small amount of space, goats and chickens are a more feasible way of producing large amounts of protein, but most of your area will be devoted to the garden. Make sure to follow these rules of thumb:

1. Use space-saving techniques.
2. Use excellent record-keeping.
3. Have a scarecrow, a deer fence, a dog, a cat, and any other garden protection you need.

The first thing you need to do is plan out how much grain you will need, and how much space it will take to grow it, and work around that. You need to know how much food your family eats, and the number of

▲ This kale was started in the greenhouse and was ready in early spring before the last frost date. It was planted very closely together to increase yield and harvested as baby greens.

plants or the area needed to produce it. You will need to plan for winter, and grow extra in case of shortages. You will also have to spend an amazing amount of time storing and preserving your food—canning, freezing, and drying take as much or more effort than growing does. A good way to estimate how much your family will eat is by looking at food storage calculations. When you know how much your whole family will eat, you will need to know how much seed is needed to grow that amount. Add the following together:

1. The amount your family eats and how much seed it will take to grow that amount.
2. The amount of seed that will be needed to grow plants next year. You need to grow extra plants that are allowed to go to seed, rather than be harvested, so you can grow more plants for the year after (if you are not buying them).
3. Overcompensate for shortages: pests, drought, storms, and disease all shrink your crop, so add one-third more to the seeds (an organic crop loses about one-third).
4. If you are feeding animals, make sure to add in seeds to grow enough food for them as well.

Increasing Yield

1. Harvest your plants every day. Broccoli, cucumber, summer squash, beans, and chard will grow more if you regularly pick them.
2. Plant your vegetables close together using companion planting and careful planning.
3. Use transplants to start plants in March so you can start harvesting in June.

4. Build a greenhouse so you can still grow food during the winter.
5. Make cold frames or plastic covers to put on top of plants in early spring.
6. Use succession planting, double-cropping (planting a new crop as soon as one is harvested), or plant every week or two.

Interplanting

Every plant has an ideal number of inches that it should be from other plants. You can grow two kinds of plants together in the same space if you carefully plan. A tall plant should be placed next to a relatively small plant that doesn't need much sun. To determine the spacing, add the spacing inches together then divide by two. For instance, to grow cabbage and turnips, add 15 plus 4 = 19. Divide 19 by 2 = 9.5. Therefore, you can plant cabbage

---

Space-Saving Ideas

1. Use plant stacking. Plants that are of different heights are planted close together utilizing the direction of the sun. For instance, the tallest plant will be placed at the end, with graduating sizes down to the front: corn, then pole beans, then kohlrabi, then onions, and then carrots.
2. Use containers on areas that have no soil (pack those plants in wherever you can), and especially use square containers because they use space better.
3. For climbing plants, use mesh for tendrils (grapes, etc.), poles for twining plants (pole beans, etc.), and lattices for other kinds of climbers.

▲ This bed is dibbled with six-inch triangles.

around nine inches from turnips. Also, remember the companion planting rules for each species.

If you have very little space, avoid summer and winter squash, cucumber, watermelon, muskmelon, cantaloupe, and corn, as they all take up lots of room, although you could try using the "bush" variety of some of these plants. If you must have squash and corn, plant the corn and the squash together in the same place. Also, you can plant pole beans in the corn and it will climb it, so you won't need poles.

More Plant Spacing

Instead of rows of crops, bed growers have the option to create equidistant planting patterns. This is much more efficient for small plants than for larger plants, but for crops like radish, carrots, and lettuce it can increase production

by 20 percent. This method is called *hexagonal* or triangle spacing because both of those shapes are used as a template.

To plan ahead for this type of spacing, you will need more seeds and seed packets. We talk a lot about standard bed sizes, and you can create your formula for your beds when you know your own standard bed size. Then, rather than using the seed packet's spacing recommendations for straight rows, you would calculate spacing based on a triangle shape. When you look at the ground, it will still look like rows, but it is offset so the next row's plant is near the center of the empty space between the plants in the row next to it. The photo shows an example of this, and the next section tells you how to calculate your seeds.

Calculating Number of Seeds per Square Foot According to Plant Spacing

	2 inches	3 inches	4 inches	6 inches	8 inches	12 inches
1 sq. ft.	40	17	10	4	3	1

Number of Seeds for Triangle Spacing per Square Foot

The seed packet will give you a spacing requirement for each plant, such as "Plant 2 inches apart." If you are using triangle spacing, this is how many seeds you will need per square foot.

The quickest, cheapest way to make a measuring stick for triangular spacing is to create a cardboard triangle with equal sides that are the same length as your spacing. For example, if you need to space your plants three inches apart, your triangle will have three-inch sides. You would place that on the soil and plant seeds at each corner. This can be created out of wood or plastic as well.

If you want a device that can make holes for you, you can create a seeding jig. This is a piece of plywood with bolts and washers through it at intervals that match your spacing. The bolts jut out of the plywood, and as you place it down flat on the soil the bolts push in, creating perfect seeding holes.

Seasonal Succession

Just like the succession growing we talked about earlier, in which plants are grown back to back to maximize space, seasonal succession is a plan to maximize your growing season. The simplest way to plan succession growing is to divide your garden space into four parts. The parts will be used for early spring, spring, summer, and fall gardens. In each of the four parts you can delay some of the planting by two weeks, so you can continuously harvest. This is often the easiest way to start out as a beginner. From there you can move on to a more complex system:

1. Divide your season into two parts—cool season and warm season. The cool season is when the night temperature is between 25°F and 60°F. The warm season is when the night temperature is 15°F cooler than the day. Normally, seed packets will recommend a planting date on or around the last frost in the spring, but some plants are hardier than that.

2. Plant cool season plants when you know that the night temperature is not below 25°F: arugula, broad beans, beets, broccoli, Brussels sprouts, cabbage, Chinese cabbage, cauliflower, collards, endive, fennel, kale, kohlrabi, lettuce, parsley, parsnip, peas, radish, spinach, and turnips can all handle cool temperatures. You may need to use the greenhouse or cold frames.

3. Plant another garden of warm season plants at the spring planting dates: artichokes, asparagus, bush beans, carrots, celery, sweet corn, cucumber, eggplant, garlic, leek, melon, onion, peanuts, sweet peppers, pole beans, potato, pumpkin, rhubarb, squash, sweet potato, Swiss chard, tomato, and watermelon.

Cold Climates

It's a challenge to grow enough food all year to feed an entire family in a cold climate. It's not impossible, just realistically very difficult. You will be able to supplement your supply with the greenhouse, but

▲ Beds covered in winter.

you'll have to strategize the rest. We do this in two ways: extending the growing season, and preserving food. To extend the growing season, we create a rotating planting season. We plant cold-hardy salad vegetables in early spring, then our summer vegetables, followed by more cold-hardy salad plants, then root crops and vegetables that can stay in the ground during the winter. In some places we can also grow a green manure crop before spring. When there is a threat of frost, some of these plants must be protected by various devices, such as a cloche, hoop house, cold frame, or greenhouse.

Carrots, turnips, leeks, other root crops, and some greens like kale are hardy to frost, so you can leave them in the ground. Cover the soil with a thick layer of hay to keep it from freezing. At the end of fall you may have a bunch of unripe green tomatoes still on the vine. To ripen those last tomatoes and save them from the frost, pull the whole plant up and hang it upside down in your basement or cellar.

Cold Climate Garden Schedule

April	June	August	September	October
Plant early garden transplants or seeds. Start transplants for the next planting.	Plant early garden transplants or seeds. Start transplants for the next planting.	Harvest spring gardens. Plant summer transplants. Start transplants for fall garden.	Harvest spring/ summer gardens. Plant fall transplants. Start transplants for winter garden.	Harvest all the gardens. Plant winter transplants.

ORCHARD

"Anyone who has a garden, park, or orchard tree has an opportunity to ensure that it offers protection, brings beauty and bears fruit for future generations. In short, every one of us should aspire to be a forester."

—*Gabriel Hemery,* The New Sylva: A Discourse of Forest and Orchard Trees for the Twenty-First Century

In the City

You are not very likely to be able to cultivate an orchard in the city, but it is included here because of the community aspect of urban farming. Orchards and food forests are an increasingly popular form of urban farming that is being embraced by cities and community centers. This is because they are not very labor-intensive and produce food for decades. If you are interested in getting city-owned land, proposing an orchard is sometimes a more appealing option.

The Perimeter

The orchard is planted at the edge of your garden, but first you have to prepare the soil and develop nitrogen-fixing leguminous plants, such as clover or some species of shrubs. Then you can plant your orchard trees in amongst the shrubs and small plants. These really shouldn't be in rows if they are for your own use, but if you are using the orchard to generate income then you should plant in rows, always making sure to form them along contours.

▲ Urban citrus trees in a village in Spain.

The Ultimate Guide to Urban Farming

Choosing the Right Trees

When choosing the right trees for your orchard, take these things into consideration:

- When the tree is mature, will it be shaped like an umbrella or be more open? Open trees let in light for intercropping.
- If you want to grow smaller trees under your larger trees, pick ones that are tolerant to shade.
- A tall tree grows very wide and will shade out undergrowth unless you want to spend the time pruning it back.
- Keep trees that don't need water away from trees that do. This will simplify the watering process.
- Some trees may stop other trees from growing well. Also, make sure to plant male and female species together for pollination.

You will also want to choose species that do well in your region. In a cold climate, apple, pear, quince, cherry, peach, plum, apricot, filbert, chestnut, walnut, hickory, olive, loquat, and pineapple guava may be grown. The species must be disease-resistant, which is more likely with a heritage variety.

Intercropping

Intercropping has been discussed previously in this book, but is just as important in the orchard. It is here in the orchard that the forest garden model has a chance to really play out. The species that you choose should be resistant to disease, won't compete with other plants for water and nutrients, and can function as a windbreak. Under the trees you can grow green manure, nitrogen-fixers, and forage crops for chickens, sheep, or pigs; repel insects; or grow grass, flowers, herbs, or vegetables until the trees get too big.

To stop grass from growing under the trees, a variety of small plants can be grown. These are only really needed in the first few years, when young trees are competing with the grass.

- Bulbs like daffodils and onion species come up in the spring and die off by summer.
- Dandelions and comfrey have deep spike roots and leaves that cover the ground.
- Fennel, dill, tansy, carrot, Queen Anne's lace, catnip, and daisy all attract wasps, bees, and friendly birds.
- Clover and leguminous plants cover the ground and create nitrogen in the soil.

Desert Orchards

When planning an orchard in a dry region, the first consideration is species. Choose trees that don't need much water, or that can withstand drought. The trees will have to be spaced farther apart than they would be in a temperate zone so that they won't compete for water, and it is a good idea to plant them during the rainy season. Trees should also be mulched. In deserts, rocks can be used to protect the roots from heat and damage, and act as a thermal mass that keeps the roots warm at night. Palm leaves or brush can be propped over the tree to protect saplings from the sun, and fencing or dogs should keep nibbling animals away. Interplant leguminous species in between.

The most efficient way to water trees in dry climates is by either using drip irrigation, which is a pipe system underground, or by building roof water and storm drains that lead into swales. The trees can be planted on the edge of the swales to take advantage of the water. If the soil is very sandy, each tree can also be planted in a hole that has been lined with mud clay so that the sand won't collapse and water will be retained.

On a slope, trees should be planted in zigzags with logs or water-runoff ditches between them. The hardiest trees should be planted at the top of the hill, with progressively less-tolerant species toward the bottom where the deepest soil and most water will be.

In a desert, you can be more flexible with your zones. If valleys and streams flow through the property, meandering through all the zones, you may need to plant trees along the pathway of the water or right next to the house to take advantage of gray water. It is much easier to grow a tree where there is water already than it is to bring water uphill to a dry place.

Pruning

Each plant is different and requires different pruning techniques and schedules, so research your tree first. Careful, balanced pruning makes your plants grow better and fruit grow bigger, while excessive pruning can kill. If cutting a diseased plant, clean your pruning shears with rubbing alcohol afterwards. Never prune the first year, and never during a drought—in fact, prune as little as possible. It is better to prune too little than too much. This is the typical pruning process:

1. Each plant is different, but in general, you can prune in early spring and late fall when all the leaves are off the tree.
2. Prune off the dead wood and the branches that are rubbing against other branches (you should only need to do to the worst ones).

3. Pinch off some of the *suckers* and *water sprouts*, but not all of them—leave a few to flower. A sucker is a small sprout that grows out of the roots of the tree and comes up out of the ground (it looks like a completely different baby tree), or right out of the trunk of the tree. A water sprout looks a lot like a sucker but is a very green shoot that comes out of branches after too much pruning or an injury. If you take them all, they will all come back, but take a few and only a few will come back.

4. You can now attempt to train the branches of the tree to be more conducive to an optimum tree shape. Fruit trees need to have more horizontal branches for easy picking. Mature fruit and nut trees need more light within the tree so they might be thinned, and they can be pruned smaller than another tree so that the branches will grow short and fat for strength and easy picking.

    Pyramid: This type of pruning keeps the tree quite a bit smaller than other methods. It is a good idea to do this type of pruning after April, rather than in the winter, to prevent silver leaf disease. It is a fairly straightforward strategy. On an established tree you would carefully cut the tree into a pyramid shape.

1. A sapling in the first year should be cut back to two feet (60 cm) high.
2. In the second year, cut off eighteen inches (45 cm) from the top of the main stem, and trim off the ends of the remaining branches to just above the previous year's healthy bud.
3. The third year, cut eighteen inches (45 cm) off the top of the main stem and trim back the ends of the top few

branches to just above the last healthy bud, or about ten inches (25 cm).

Silver leaf disease: Found all over the world, *Chondrostereum purpureum* is a fungal disease that attacks just about any deciduous tree. Any home orchard is susceptible to it, and it can spread between species through wounds in the bark. It is recognizable by the silvery sheen it makes on the leaves when it damages the leaf cells. Apples tend to be fairly hardy and can usually recover, but other species can die from it. In the meantime, it reduces the amount and quality of the fruit you do get. To prevent silver leaf, it is important to follow smart pruning strategies.

- Use sharp, high-quality pruning shears so that the cuts you make are clean and smooth, and so the wood won't split. If the branch is big, use a saw to make a clean cut.
- Prune in the spring on a warm sunny day, not in the winter. Silver leaf prefers cool, wet conditions.
- If you are going to use wound dressing after the cut dries, do it on the same day. The dressing should be applied thickly with several coats. You should also know that pruning paint may be expensive, but making your own is ineffective. If you can't buy some, don't use any. The tree's natural defenses will help it heal and homemade stuff will actually prevent it from healing.

Bush: Also called *open-centered*, the bush shape has a stem of two and a half feet (0.8 m). The aim is to create a shape that has (obviously) an open center. This is done in early spring.

1. On a tree that is grown, pinch off any buds on the bottom of the trunk and pull off any suckers.

2. This type of pruning is not very extensive. You should only have to remove any stems and branches that are crossing or vertical, and any weak or diseased ones.

3. If you need to thin the branches more later, you can do so in July.

Fan: Fan training is used when a tree is grown up against a wall or fence at least six feet high, and hopefully facing the south or southwest for most varieties. Dwarf varieties work the best for this.

1. The tree should be planted between six and nine inches (15–22 cm) away from the wall, angled slightly toward it.

2. It is best to do the cutting in the spring, when branches can heal quickly. In the first year, when a sapling has no branches, cut back the main stem to fifteen inches (38 cm), making sure that there are at least three strong buds.

3. In the summer, put two stakes into the ground at forty-five degrees on either side of the tree and tie the two side branches to them to start creating the fan shape.

4. Repeat the next year for two more inside branches at a lesser angle.

5. When a tree is already a couple of years old and hasn't been trained, you will have to cut back the main stem to about fifteen inches (38 cm), put in the stakes, and cut each arm of the fan by two-thirds to just above an upward-facing bud.

6. At this point, the tree will have two arms extending from each side. In the summer, tie four shoots from each arm thirty degrees from the main arm so that they will create a fan shape.

7. Pinch off any shoots that are growing out toward the wall. Cut back any other shoots so that there is only one leaf left.

8. The next spring, cut back the four branches on each side by one-third, just above an upward-facing bud if you can.

Planting Trees

1. A tree has a *root ball*: a ball of soil that is surrounding the mass of roots. Don't let this ball dry out before you plant it. If the ball is wrapped in something, take that wrapping off before you plant.
2. Dig a hole deep enough that the root ball will be covered by the soil. A tree has a part at the base where it begins to flare out and turn into roots. This flare is where you will want your hole to come to.
3. Put the tree in the hole and fill it with dirt. Leave a circular compression or trench all around the tree a little thinner than the root ball so that water can settle in it.
4. Spread fully composted organic material on top of the soil all around the tree (don't mix it with the soil because it can actually cause harm), and water very thoroughly to soak the roots. Don't step on the dirt around the tree.
5. You shouldn't have to stake a tree unless the dirt around it is loose—trees need to learn to hold themselves up. Pruning also does not help the tree grow unless you are only removing broken branches.
6. When watering, soak down to the roots, but then wait a while until the soil dries out to water again. Only swamp trees like their roots wet all the time.

Grafting

Grafting is when a branch from one fruit tree is grafted onto another of a different type. The purpose of this is varied, but the most common reason is that putting a difficult-to-grow fruit species on an easy-to-grow tree will hopefully make it grow faster and easier. Grafting makes it possible to grow several kinds of apples on the same tree, saving you space. However, you can only really graft a cherry tree onto a cherry tree, an apple tree to an apple tree, etc.

1. In late winter (February), cut the scions (fruit-tree shoots with buds) that you want to graft at both ends so they are about two inches long and as thick as a pencil. Do this by pressing them against the knife with your thumb and rolling them back and forth.
2. Start grafting in late March or April, when the buds on the trees you are grafting start to swell. Bring grafting wax with you in a dish of warm water.
3. On the tree you are grafting to, saw the top off a branch and split the branch down the middle with a chisel or knife. The split can be small. Wedge the split open a little bit with a small piece of wood.
4. Shape the end of two scions so that they will wedge into the branch better.
5. Carefully put two scions into the cleft, one on each side, so that the end of the branch appears to be forked. Take out the wedge.
6. Take a ball of grafting wax, lay it on the end of the stalk between the two scions, and press it down so that it seals up all the openings in the wood and is firmly in place. Grafting wax is any sealant to be used for plugging plant wounds. Some people use simple Elmer's glue, but beeswax or other sealants will also work. Bind it with tape or cloth.

ORGANIC CONTROL METHODS

"Growing up, my mom made dinner every night. Usually this would be a large salad with kale, carrots, tomatoes, cucumbers, all organic, of course, and sometimes she'd sprinkle nuts on top for texture. Kale has a metallic taste, like chewing on the hood of a Mercedes. No, something safer, like a Volvo."

—*Jarod Kintz, Gosh,* I probably shouldn't publish this.

The Definition of Organic

Early in 1940, English botanist Sir Albert Howard wrote *An Agricultural Testament,* describing alternative organic methods for building soil fertility and health. This was in direct conflict with the respected German chemist Justus von Liebig, who had set the modern scientific standard in plant nutrition a hundred years earlier and is considered the father of the modern fertilizer industry. He promoted the idea that plants need only three major components: nitrogen, phosphorous, and potassium (symbolized by their chemical letters NPK), give or take some trace minerals. Among his many accomplishments were the invention of nitrogen-based fertilizer and Oxo beef bullion.

Von Liebig declared that humus was not important, and what plants really needed was intense applications of NPK. Sir Howard, on the other hand, created the Law of Return, which was almost the

▲ Organic growing is labor-intensive in that it requires a lot of work by hand.

complete opposite: whatever comes from the soil must be returned to it. It was an ideology that flew in the face of the Industrial Revolution because it seemed to turn against science and progress. Because of this, Howard saw little credit for his ideas, and he remained obscure for the rest of his life. A young man named Jerome (J. I.) Rodale read Howard's book *The Soil and Health*, and it inspired him so much that it became his life's work. He became a dedicated activist, promoting organic growing in his magazine and eventually meeting Howard and inviting him to write some articles. Jerome Rodale became the ambassador for change in the 1930s.

These two men defined *organic* as we know it today. Conventional methods of fertilizing soil today are exactly the same as those invented during the 1840s: adding caustic and potent NPK in its pure forms, thereby killing the soil in the process. The living matter that makes up healthy loam quickly dies under repeated applications. Organic growing depends only on the natural waste that comes from the earth to feed plants: the roots and stems left over from harvest, manure from animals, vegetable matter that begins to rot—all the components of a compost pile. Rather than growing bigger plants faster, organic is content with building a higher yield over time through better soil.

For the first few years of his magazine, Rodale targeted farmers, trying to educate them on organic farming. Many hours and wasted dollars later, he realized that most farmers at that time truly didn't care, and

> **Organic:** Building soil health by returning to the soil what the soil produces.

he abandoned the effort. Instead, he turned to the home gardener and the consumer, who began to really drive the movement forward. It wasn't until the early 1950s, however, when a congressman became sick after exposure to DDT that the term *organic* began to take off as a household word. Congressman James Delaney launched the first government investigation into chemicals used to produce food, and the few organic farmers who did exist at the time wholeheartedly supported his sudden interest. Delaney invited J. I. Rodale to speak to Congress, but he came across as a romantic. In turn, Rodale felt that the government and the USDA were ignorant and was disgusted with their lack of interest in researching the matter. He was attacked by public relations campaigns on all sides, including Monsanto, who published a propaganda pamphlet with pictures showing organically grown fields of corn yielding far less than those grown with chemicals.

In North America, the organic movement very quickly became political. Mistakenly, it was defined by the idea that people must rebel against the Industrial Revolution in order to be on par with growing organically. It was seen as a step backwards.

Not much has changed in North America. Today, *organic* can mean any number of things as defined by the USDA organic standards, which were eventually created by the government. However, in its pure form, true organic means caring for the soil, not the plants. Organic growers feed the soil what came from the soil, and conventional growers feed plants lab-created minerals.

We know the basic principles behind organic growing are the soil, composting, and seeds. But, there are a few more strategies to help you be successful: placement, protection, and prevention.

Placement

The placement of crops in organic growing is one of the most important things to consider. It involves knowing the lay of the land and what else is growing nearby. Each plant has its own specific needs for soil type, water, sun, and temperature, and the environment you provide can mean the difference between a healthy plant and no plant at all. So, the first thing to do is map out your garden space. What are your lowest and highest temperatures every year? What kinds of soils do you have? Where do you get the most sun?

Raised bed or container gardening will be vastly more successful for you than tilling and row-cropping. Raised beds prevent all kinds of pests from getting access to your plants, help with drainage, and are perfect for no-till soil care. Container gardening can be even better because it deters all kinds of burrowing beetles and creatures, and you can move the containers around year to year to confuse other pests.

Once you've placed your beds in the warmest, sunniest spot and filled them with perfect soil, you now have to decide on the placement of each plant species. This is where *crop rotation* comes in. Crop rotation is not succession planting, which has the goal of increasing production by growing more plants in one place. Instead, crop rotation is a strategy of moving plants around to make sure the soil does not get tired out or depleted, and to confuse pests who tend to return to the same spot every year. This plan always includes a legume

that returns nitrogen to the soil, and big farms will often rotate alfalfa or clover, but you can use peas or beans as well. So, every year, at least once, it's a good idea to grow a legume in every square inch of soil. By returning nitrogen to the soil, the legume is doing a job that would require lots of fertilizer instead, but it does this without energy or cost, and you might get some food out of it as well.

The next part of crop rotation requires an understanding of some of the basic vegetable families. The "Everything else" category includes plants that don't belong in a whole family, but have their own rotation.

Alliums: onion and garlic family

Brassicas: cabbage family, including Brussels sprouts, kohlrabi, kale, pak choi, arugula, turnips, and rutabagas

Chenopodiacaea: beet family, including Swiss chard and spinach

Cucurbits: cucumber, melon, and squash family

Legumes: peas and beans

Solanaceae: tomatoes, peppers, potatoes, and eggplants

Umbeliferae: carrot family, including cilantro, dill, parsley, parsnip, and fennel

Everything else: mint, oregano, rosemary, sage, basil, berry fruit, lettuce, endive, cress, Jerusalem artichoke, corn, and asparagus

Imagine you have a little garden bed. In that bed you decide to grow spinach, which takes only a couple of months to grow. Once it's done, what do you grow next? Spinach does not feed heavily from the soil and is not prone to many problems, so you can follow it up with most other plants. You can't grow beets or Swiss chard because they are from the same family, so you need to choose from another family.

You predict that you can grow at least one more crop in the heat of summer before cold weather hits. It's a good idea to grow brassicas right after growing peas because they need a lot of nitrogen, so brassicas aren't a good choice after spinach. Carrots prefer cooler weather and it will still be too warm for them, so some good options to follow the spinach crop might be bunching onions, squash, or even tomatoes seedlings depending on the length of the growing season. Then you can sow peas at the end of the year. It's important to also add an inch of rich compost between each planting. For tomatoes, you can add a little more.

Most of us grow more than one type of vegetable at a time, so this is where companion planting comes in, which we discussed in previous sections. Some people really take this to extremes by trying to follow companion planting tables religiously and making very complicated growing patterns, but just understanding plant families and knowing a few rules is enough. It's safe to say that a plant can grow next to one from its own family, but it's also a good strategy to space them out, if possible, because they are susceptible to the same pests. If you move them away from each other one is more likely to survive.

These companion plants can be intercropped. For example, in the crop rotation we talked about earlier, the spinach can be planted with radishes in between, and the bunching onions that follow could eventually have carrots among them as well. What was originally one crop of spinach becomes radishes, carrots, onions, peas, *and* spinach. Putting all of these together in the same bed in the same year not only increases our chances

of getting a good crop, but increases our yields overall.

Protection

Most professional organic growers use man-made materials to protect their crops, whether that is a massive commercial greenhouse or a plastic sheet thrown on the ground. I hate plastic as much as the next person, but it is a tremendously useful and necessary tool in organic growing. There are now biodegradable plastic sheet mulches available. Most people in North America live in a temperate region, with cold, wet winters and hot, dry summers. This climate also provides the perfect home for a variety of molds and pests that don't exist in the desert. We may think people in the South are the only ones who can grow all year-round, but that's not true. Welcome to the miracle of plastic.

There are other ways we can protect our crops, but they all use similar materials. Glass is an obvious choice, but unless you have a bunch of old windows, it is much more expensive. You are more likely to be able to get plastic, and there are two kinds. The first is clear and lightweight and should be three micrometers thick (called 3-mil poly at the store). These sheets are for use as "floating row covers," which is just a fancy phrase for the plastic thrown on top of plants at night to prevent frost damage. You can use this cheap stuff to lengthen your growing season by about a month in northern regions, and several months in southern regions. The second type is clear and heavyweight and should be a minimum of 6-mil poly (six micrometers). This plastic sheeting is used with hoops made out of electrical conduit or PVC, which can be built as a low tunnel on each raised bed, or as a big

▲ Peppers and tomatoes grown in a greenhouse equipped with extra lighting and ventilation.

high tunnel or greenhouse out of wood or metal. Depending on the weather, the 6-mil sheeting can be reused a second year, but the 3-mil can't. The heavier plastic can allow you to grow all your crops much earlier in the year. Plastic floating row covers can only help you grow cool-weather crops earlier. They won't protect your plants from snow. For example, you can use the 3-mil floating cover to grow kale into the winter, but with a 6-mil tunnel you can grow carrots, broccoli, and other cool-weather crops in the dead of winter.

The other kind of plastic cover is nonwoven polypropylene. It looks like fabric and provides the same frost and pest protection and temperature control as plastic, but it allows moisture and light in, and it breathes better. This sheeting is used extensively by northern growers and

tends to be more expensive than plastic but is supposed to last longer. It's a good idea to try both in your area and see which you like better. This fabric cover can be used earlier as a low hoop tunnel or as a floating row cover.

The final material is mulch. Organic certification stipulates that you cannot use plastic mulch permanently, because these big plastic sheets begin to degrade over time and add little bits of plastic to the soil. Plastic mulch keeps the weeds down exponentially, reducing labor dramatically, but you have to pull it up every year at least once and do a proper crop rotation. If you do that, plastic mulch can be extremely valuable. The black color warms up the soil earlier, keeps moisture in—which reduces your watering time—and deters burrowing pests. Alternatively, you could use leaves, and bark works very well for pathways. Leaf mulch has the added value of not having to be removed and provides more needed nutrients to the soil, but it isn't effective at deterring pests and doesn't stop weeds as well.

It is likely that as you start out, you may have to prioritize what is protected based on the size of your budget. Your warm weather crops like tomatoes, peppers, and cucumbers need heat in the beginning of the year, and a greenhouse can give it to them. Greens are another crop that is a priority, especially from the brassica family. They are extremely susceptible to pests that hold over from year to year, but if you tuck them in tightly with a fiber floating row cover, you can keep the pests at bay.

Strategies for plant disease prevention:

1. Rotate crops—don't plant the same thing in the same spot two years in a row.
2. Don't work in the garden when it's wet.
3. If the plant is diseased, don't save seed.
4. Clean up old debris.
5. Maintain even soil moisture.
6. Fertilize properly—don't overdo it.
7. Don't put diseased plants in the compost pile.
8. Try not to get the tops of the plants wet when watering.
9. Wash your hands after touching a diseased plant.
10. Grow resistant varieties.
11. Prune and weed to improve air circulation.
12. Avoid injuring plants and trees.
13. Sterilize pruning shears after pruning diseased plants.

Prevention

This is just a reiteration of everything we have talked about regarding organic growing. It starts with composting religiously and repeatedly, adding it to the soil every time you grow something. Simply paying attention to what is going on in the dirt throughout the process will really help you.

Placement and protection are both part of prevention. Prevention is much easier than repair. Once attacked by a pest, a plant has a very difficult time recovering. Some of the energy that would have gone to producing an edible fruit or vegetable will now have to go to repair. If you are fraught with fungus and disease, you will have to destroy every plant that is even slightly infected in order to save the rest. And you can't throw them in your compost pile. It means making hard choices, and even when things look like the end of the world, being resolved to never use chemicals. Without step-by-step prevention, organic growing doesn't work.

Weeds are a big problem in the spring. Your tiny vegetable seedlings begin to grow only to be choked out by weeds that grow much faster. Most plants can be transplanted, which not only starts them off earlier, but also gives them a jump on the weeds. Beans, corn, cucurbits, mustard, peas, and turnips are the few plants that are not worth transplanting, but everything else can be planted in flats or trays according to the directions on the packet.

If you have prepared the soil properly, it will be full of compost and should sit at a nice, even 6.0 or 6.5 on the pH scale, with not too much of anything. However, some plants like broccoli and tomatoes use so many nutrients during the season that

▲ Plastic mulch keeps lettuce free of weeds, which can create a lot of work for the farmer in growing and especially harvesting.

they need a kick. Kelp tea is the easiest way to do this. You can buy liquid kelp, or make your own tea by buying kelp meal or gathering fresh kelp from the ocean. Soak the kelp in water for a couple of weeks and pour it directly on the plants. Neem and alfalfa powder can be mixed in for an even more potent tea. This will boost your plants' vitality and help keep their immune systems strong.

Even if you decide that you can't grow very much food, knowing how food is grown organically will help you to know your farmer's practices better and make better choices. Your best defense is knowledge.

Plants to Grow to Attract Good Bugs or Repel Bad Ones

Attract Good Insects	
Bees	bee balm, borage, summer savory
Hoverflies	buckwheat, German chamomile, dill, morning glory, parsley
Ladybugs	yarrow
Predatory ground beetles	amaranth, lovage
Predatory wasps	anise, borage, German chamomile, dill, yarrow
Deter Bad Insects	
Aphids	anise, catnip, chervil, coriander, dill, garlic, mint, yellow nasturtiums, peppermint, petunias, sunflowers
Ants	catnip, mint, tansy, pennyroyal
Asparagus beetle	parsley, petunias, pot marigold
Bean beetle	rosemary, tomato, summer savory
Black flea beetle	sage, catnip, wormwood
Blister beetle	horseradish
Cabbage moth	hyssop, peppermint, rosemary, sage, summer savory
Cabbage worm	borage, thyme
Carrot rust flies	rosemary, salsify, wormwood, sage
Codling moths	garlic, mint
Colorado potato bug	flax, horseradish, eggplant
Corn borer	radish
Cucumber beetle	nasturtiums, radish
Flea beetle	catnip, hyssop, mint, peppermint
Fleas	lavender, pennyroyal
Flies	basil, rue
Gophers	elderberry
Japanese beetle	catnip, chives, white flowering chrysanthemum, garlic, rue, tansy, white geraniums, larkspur
Leafhopper	petunias
Mexican flea beetle	petunias
Mosquitoes	basil
Moth	costmary, lavender
Potato beetle	coriander, horseradish
Root maggot	garlic
Root nematodes	chrysanthemum

Slugs	comfrey, wormwood
Snails	garlic
Spider mites	coriander, dill
Striped cucumber beetle	tansy
Striped pumpkin beetles	nasturtium
Squash vine borer	radish
Squash bug	catnip, dill, nasturtiums, tansy
Thrips	basil
Tomato hornworm	borage, petunias, pot marigold
Weevil	catnip
White cabbage butterfly	peppermint, wormwood
White fly	nasturtiums, marigolds (calendula)
Wooly aphid	clover

Plant pest-control strategies:
1. Turn over your soil in the fall so the birds can eat the bugs.
2. Rotate where you plant your crops every year, so bugs can't get a residence.
3. Fertilize and help your soil.
4. If you can't weed, mulch.
5. Keep your garden clear of weeds and debris.
6. Cultivate the plants in the fall.
7. Destroy infected plants.
8. Keep the ground clear of everything around trees.

Organic Pest Control Methods

Note: Insects usually go through several life stages: egg, pupae, larva, and adult. A pupae is like a cocoon, while a larva looks more like a grub.

Pest	Control method
Alfalfa weevil	Encourage parasitic wasps.
Aphids	Spray with watered-down clay or soapy water, and bring in ladybugs.
Asparagus beetle	Shake beetles into can of soapy water, or spray tea on plants.
Black flea beetle	Dust with soot and ashes, plant near shade, and spray with garlic or hot pepper.
Blister beetle	Also toxic to horses or grazing animals. Cut hay with a sickle bar or rotary mower, and don't crimp or crush hay.

Pest	Control method
Cabbage looper	Sprinkle worms with flour or salt.
Cabbage maggot	Apply wood ashes to the soil.
Cabbage worm	Put sour milk in the center of the cabbage head, and dust with 1 cup flour. Or use mint cuttings as mulch.
Carrot rust fly	Sprinkle wood ashes at plant base.
Codling moth	Spray with soapy water.
Colorado potato beetle	When plants are wet, spray with wheat bran, remove beetles and eggs by hand, and spray with mix of basil leaves and water. Cover ground with 1 inch of clean hay or straw mulch.
Corn earworm	Put a little mineral oil in the silk on the tip of each ear.
Cutworm	Sprinkle wood ashes around plant and press a tin can—with the bottom cut out—around the stem 3 inches deep.
Harlequin bug	Handpick bugs off.
Hessian fly	Encourage parasitic wasps.
Hornworm	Handpick off and sprinkle dried hot peppers on plant.
Leafhopper	Shelter plants.
Mexican bean beetle	Spray with garlic, destroy eggs, handpick beetles, and plant earlier.
Mosquito	Empty out all standing water.
Moth	Sprinkle around dried sprigs of lavender.
Onion maggot	Plant onions all over instead of in one place.
Slug	Handpick off, make borders of ashes or sand, and mulch with wood shavings or oak leaves.
Spider mite	Spray cold water, or spray mix of wheat flour, buttermilk, and water on leaves. Or spray soapy water or spray coriander, and introduce predators.
Squash bug	Handpick, grow on trellises, and dust with wood ashes.
Squash vine borer	Pile up soil as high as the blossoms.
Striped cucumber beetle	Handpick, mulch heavily, dust with wood ashes, and grow with trellises.
Tarnished plant bug	Remove plant after harvest.
Thrips	Spray with oil and water mix or spray with soapy water.
Weevil	Hill up soil around sweet potato vines.
Wireworm	Plant green manure like clover.

Natural Pest Control Concoctions

The following is a list of homemade mixtures that can prevent and deter pests.

Bay leaf	Sprinkle dried leaves on plants.
Catnip	Spread sprigs around and it will repel a variety of pests.
Chrysanthemum (pyrethrum)	Mix with water and spray as a general insecticide.
Garlic	Concentrated garlic spray deters fungus and insects.
Horseradish	The root can be made into a spray, either raw or as a tea.
Hot peppers	Make a spray from the tea. It repels most insects.
Kelp	Use as powder or tea as spray to kill bugs. It also fertilizes.
Lemon balm	Use as powder and sprinkle throughout garden.
Marigold	Plant the scented varieties throughout the garden.
Petunia	The leaves can make a tea for a potent bug spray.

Pest-Eating Creatures Helpful to Farms

The following are beneficial creatures that should be encouraged to take up residence in your farm.

Spiders	Do not kill the spiders you find in your garden. They eat all kinds of pests!
Minute pirate bug	Eats spider mites.
Mite destroyer beetle	Eats spider mites, introduce into garden.
Parasitic wasp	Eats Hessian flies and others.
Six-spotted thrips	Eats spider mites, introduce into the garden.
Toad	Lay a flower pot upsidedown in your garden.
Purple martin birds	Make a birdhouse with many apartments or a group of birdhouses clustered together, painted white, and put on a high pole away from trees and buildings. Remove other bird's nests except the purple martins, and take down the house in the winter.

Poisonous Plants as Natural Pesticides

Note: These plants are poisonous to humans.

Plant	Kills or Prevents	Notes
Four-o'clocks	Kills Japanese beetles	Poisonous especially to children.
Larkspur	Repels aphids	Poisonous to both humans and animals.
Marigold (Calendula)	Repels insects, kills nematodes	Unlike pot marigold, French and Mexican marigolds are poisonous.
Mole plants	Repels rodents	Isn't good for humans, also gets out of control.
Stinging nettles	Repels mites and aphids	Hairs on the leaves have formic acid which stings you. However, these can be ingested when properly prepared.
Tansy	Repels nettles, beetles, squash bugs, cutworms	Toxic to many animals. Don't let it go to seed or it will grow everywhere.
Wormwood	Repels moths and most other insects	This medicinal herb is used to create a botanical poison that should not be used directly on food crops.

Crop Rotation

It is a little overwhelming to look through all of these tables of companion plants, succession tables, and plant families and wonder what you should be planting next. The table below is a simple guide to rotating crops in order to prevent insects and disease and maintain your soil quality.

Current Plant	To Plant Next
Beans	Cauliflower, carrots, broccoli, cabbage, corn; not onions or garlic
Beets	Spinach
Carrots	Lettuce, tomatoes; not dill
Cucumbers	Peas, radishes; not potatoes
Kale	Beans, peas
Lettuce	Carrots, cucumbers, radishes
Onions	Radish, lettuce; not beans
Peas	Carrots, beans, corn
Potatoes	Beans, cabbage, corn, turnips; not tomatoes, squash, or pumpkins
Radishes	Beans
Tomatoes	Carrots, onions

Identifying diseases:

A disease often looks like a weird brown or rust spot on some part of the plant. *Tan spot wheat disease* makes small brown dots on the leaves, while *alfalfa phytophthora root rot* makes a big brown spot on the root. Another kind of disease is a caused by a nematode, or a worm that creates cysts on the plants. The cysts make the plant stunted and/or yellow. There is no cure for this disease except for prevention and removal. Every location has different diseases, and the best possible identification method is taking the plant to your local extension office.

Infected plants:

1. Destroy infected plants/trees or pruned-off parts.
2. Scrape off loose bark from trees.
3. Cut out diseased wood, then patch with tree-patching compound.

PROCESSING

"For now, the corn house fill'd, the harvest home, The invited neighbours to the husking come; a frolic scene, where work, and mirth, and play, Unite their charms to cheer the hours away."
—*Joel Barlow,* The Hasty Pudding

The Plants section (Chapter 4) contains information on best practices for harvesting each individual type of crop. This section tells you what you can do after you have harvested your crops.

The Harvest Station

Harvesting and processing is your biggest task as a farmer, so a well-designed processing station is invaluable. It should be located under cover as close as possible to your growing space. You will need the following:

- two sturdy tables, one for dirty produce and one for clean produce

▲ This is an ideal setup for processing: under cover, with a steel sink basin and drying area. Note: the farmer does not have to bend over.

- a hose hooked to a sprayer for washing produce
- ideally, a large double-basin sink. This can be an old restaurant sink, two bathtubs, or even large plastic bins
- a way to catch the runoff from your washing station and divert it back to the garden

You will also need a variety of tools depending on what you are harvesting. If you are packaging herbs or salads, you will also need plastic bags, berry baskets, or other food-safe packaging. You will definitely need a durable salad spinner if you are packaging mixed greens or *mesclun mix*. If you are lucky, you can get an industrial-sized one from a restaurant sale. Or, perhaps even better, you can use an old washing machine set on the spin cycle. The key to protecting the leaves from damage is not overloading the machine.

Food Safety

Produce that is straight from the ground is incredibly dirty. Organic growers run the risk, as frequent users of manure, that their produce will harbor dangerous *E. coli* and other nasty bacteria. This is not only a health risk, it's a liability risk for you. This means using good hygiene as a rule throughout every part of your farm operation. Make sure the manure you use is fully composted and don't let any animals poop in your beds.

After harvesting the produce, keep it separate from the clean produce and the processing area. This is why having two tables is such a good idea. You can bring your bins or harvest trays to the table, cut off tops or pick out the bad stuff, which can be recycled into the compost.

Now it's time to wash the produce, the harvest area, and your hands. There are a variety of strategies for washing produce to preserve quality and save time. This is detailed under each plant in the plant guide, but most of the time you will use a solution of 100 parts per million of chlorine bleach in clean water. This is the same as ½ teaspoon per liter. To do this, first wash your hands and the processing area with the soap, and scrub most of the dirt off the produce. Wash your hands again and clean the area again, this time sanitizing the sinks, tables, and tools with the bleach solution. Wash the produce with the bleach solution and set aside on the clean produce table. Sanitize your salad spinner before using it as well.

Farm refrigerators also get very dirty, whether it's your home fridge or a dedicated farm fridge/walk-in cooler you use before the farmers' market. It's a good idea to start your processing ritual by cleaning out the fridge, using the bleach formula to sanitize it. Get all the dirt out so you can start fresh each week.

Freezer Box Fridge

A practical alternative to a regular fridge is difficult to find. Even the most energy-efficient models are woeful energy hogs, and yet an effective method of keeping all of your produce cool is so necessary, especially when you are growing so much fresh food. A regular home-sized refrigerator is not large enough for a profitable farm, so you will need at least two or three, which cost a lot to run.

The simplest and easiest solution is to take a chest freezer and replace the thermostat with a regular refrigerator thermostat. Cold air is heavier and tends to fall, which means that every time you open your typical vertical fridge, all the cold air falls right out the door. A chest freezer has the door on the top to prevent this inefficient escape of cold air. These kinds of cheap chest freezers can be found on Craigslist or other websites, and many farms just find multiple old fridges and run several at a time. The freezer box fridge is just slightly more efficient than an upright fridge.

Walk-In Storage Cooler

Ideally, a serious farmer will want and need a walk-in cooler. This is basically a small room that acts as a refrigerator and creates a perfect temperature and humidity level for keeping vegetables fresh. The simplest and cheapest solution is to frame out the corner of a room and insulate it with pink foam insulation (at least a couple inches thick). Insert an air conditioning unit into the wall and use a CoolBot device to control the airconditioning unit to keep the room at a cold temperature. While a commercial cooler can cost many thousands of dollars, a CoolBot room costs around $500 or less.

Freezing

The best temperature for freezing vegetables is –5°F, but to save energy you can go as high as 0°F (but no higher!). Put cartons or buckets full of water into the bottom of your freezer, so if the electricity goes out, food will last longer, and you will have a small water supply. Keep the freezer in the coolest room of the house, but not where it freezes, since it can withstand hot temperatures but not cold.

Before you can freeze vegetables, you must *blanch* them. Blanching slows or stops the enzymes that make vegetables lose their flavor and color. There are two ways to do this: boiling and steaming. If you blanch too much they will lose nutritional

| Food that does better if you freeze it than drying or canning: |
Asparagus, sweet green peas, snow peas, whole berries, melons, spinach, kale, broccoli, cauliflower, freshwater fish

Food that you can't refreeze after thawing:
Any kind of meat, ice cream, and vegetables. Fruit and bread you can refreeze without any problems.

value, but blanching too little will speed up the enzyme breakdown and make them all brown and wilted. You have to stick to tried-and-true blanching times to do this right.

The day before you start, turn the freezer temperature down to –10°F (–23°C), which will help everything to freeze quickly. Be ready to label every item with the date and what it is. Don't forget to turn the temperature back to 0°F (–17°C) when everything is frozen. Frozen fruits and vegetables will last about a year, with the exception of onions, which last about eight months, and baked foods can last six months. Animal products and meat only last three to six months.

To boil vegetables, wash them thoroughly and drain well. Some foods can be frozen whole; check the table for those. Most must be trimmed and chopped. You will need one gallon (3.7 L) of water per pound (0.5 kg) of prepared vegetables, or two gallons (7.6 L) per pound (0.5 kg) of leafy greens. Bring the water to a boil, and lower the food in with a wire basket, mesh bag, or metal strainer, which will slow down the boil. It should take less than a minute for the water to get back up to a boil; if it takes more, you are using too much water.

Keep them submerged for the time specified on the chart, then pull them out and quickly chill them in ice water for the same length of time that you boiled them. Drain them well, pack into a container or a zip-up freezer bag, with as little air as possible, and put them in the freezer.

Steaming is almost exactly the same, but instead of submersing foods, use only about two inches (5 cm) of boiling water in the pot. The steamer basket is lowered in, and the lid is put on. You start timing it as soon as the steam starts trying to push out of the lid again.

Onions, peppers, and herbs don't need to be blanched at all. Squash, pumpkins, sweet potatoes, and beets need to be fully cooked.

Blanching Chart

Food Type	Steaming	Boiling
Artichoke	hearts: 8 minutes	hearts: 7 minutes
Asparagus	small stalk: 2 minutes medium stalk: 3 minutes large stalk: 3 minutes	medium stalk: 3 minutes
Bamboo shoots		10 minutes
Bean sprouts		5 minutes
Beet greens		2½ minutes
Black-eyed peas	2½ minutes	2 minutes
Broad bean pods		4 minutes
Broccoli	5 minutes	
Brussels sprouts	small heads: 3 minutes medium heads: 4 minutes large heads: 5 minutes	
Butter beans	small: 2 minutes medium: 3 minutes large: 4 minutes	
Cabbage	shredded: 2 minutes wedges: 3 minutes	1½ minutes
Carrots	whole: 5 minutes diced/sliced: 3½ minutes	sliced: 3 minutes
Cauliflower	3½ minutes	3 minutes
Celery	diced: 3½ minutes	diced: 3 minutes
Chard		2½ minutes
Chayote	diced: 2½ minutes	diced: 2 minutes
Chinese cabbage		shredded: 1½ minutes

Food Type	Steaming	Boiling
Collard greens	3 minutes	2½ minutes
Corn on the cob	small ears: 7 minutes medium ears: 9 minutes large ears: 11 minutes (note: cooling time double)	small ears: 6 minutes medium ears: 9 minutes large ears: 10 minutes (note: cooling time double)
Corn cut from the cob	5 minutes	4 minutes
Dasheen	3 minutes	2½ minutes
Eggplant	1½-inch slices: 4½ minutes	1½-inch slices: 4 minutes
Green beans	3 minutes	2½ minutes
Green peas	2 minutes	
Greens	2 minutes	
Irish potatoes	4 minutes	
Jerusalem artichokes	4 minutes	
Kale		2½ minutes
Kohlrabi	whole: 3 minutes diced: 1¾ minute	diced: 1 minute
Lima beans	small: 2 minutes medium: 3 minutes large: 4 minutes	small: 1½ minutes medium: 2½ minutes large: 3½ minutes
Mushrooms	whole: 5 minutes sliced: 4 minutes buttons/quarters: 3½ minutes	medium, whole: 5 minutes
Mustard greens		2½ minutes
Okra	small pods: 3 minutes medium pods: 4 minutes	small pods: 3 minutes medium: 4 minutes
Parsnip	3 minutes	2 minutes
Peas	edible pod: 2 minutes	edible pod: 1½ minutes
Pinto beans	small: 2 minutes medium: 3 minutes large: 4 minutes	
Rutabagas	diced: 2½ minutes	diced: 2 minutes
Shell beans		1¾ minutes
Snap beans	3 minutes	
Soybeans	in pod: 3 minutes	4 minutes
Spinach		2½ minutes
Summer squash	2 minutes	

Food Type	Steaming	Boiling
Sweet peppers	halves: 3 minutes strips/rings: 2 minutes	
Turnips	diced: 2½ minutes	diced: 2 minutes
Turnip greens		2½ minutes
Wax beans	2½ minutes	3 minutes

Drying

Air drying works well for many herbs as well as alliums like onions and garlic. Herbs can be tied in bunches, and onions and garlic can be braided together by their tops. They are simply hung upside down from the ceiling. Some people do this in the greenhouse, and others recommend a cool, dark place. Your cold storage downstairs may be too humid, but a large cool pantry or even a big cupboard will work. It should take about two weeks to completely dry, and they will hold their flavor longer if you just keep them there. Just take some as you need them.

▲ I have used round dehydrators, but for a farmer, the large rectangular Excaliber dehydrator is a necessity.

The Ultimate Guide to Urban Farming

Sun drying is a no-cost, low-energy dehydration method that uses the power of the sun. This method works well in a hot, dry climate. In temperate regions, a large reflector can be made to help focus the sun's rays, like a big solar cooker, but it will still take a few days if it is successful at all. The alternative is an electric dehydrator. There are many on the market today that use very little energy to power a heating element and fan, and there are some that work effectively with a fan alone. They can dry food of any kind (including chips, jerky, and fruit leather); they can make stale chips and crackers taste better; de-crystalize honey; dry bread sticks, pasta, flowers, and dyed wool; start seedlings; grow sprouts; and make yogurt. A low-wattage dehydrator can run well on a home power source.

The process of preparing food for dehydration is a little time-consuming but well worth the effort. The most important part is deciding when the dehydration is done. Follow this list for prepping and drying fruit, vegetables, and meat:

1. Use only ripe fruits and vegetables. Wash everything thoroughly, peel, and slice them very thin unless you are dehydrating peas or corn, which can just be removed from pods and cobs. For seed pods, harvest them before they burst and use them whole. For meat, choose lean cuts of beef, buffalo, goat, or deer. Don't use pork, which is too fatty. Cut into trimmed strips around an inch wide and half an inch tall, and as long as you want, cut along the grain. Sprinkle with ground pepper and salt. This process is the most time-consuming step. There are simple devices such as apple corers and peelers that can help speed this process along. Overripe or even fermenting fruit can still make good fruit leather. Wash, peel, remove seeds and pits, and then grind it up by mashing or blending. The puree must be thin enough to pour but not watery. If it is too thick, add fruit juice or water; if it is too thin, add another kind of fruit puree.

2. Normally, when you dehydrate foods, you want to soak fruits and vegetables in a solution of vitamin C or sugar for five minutes to prevent oxidation and discoloration. However, this extra soaking will make it difficult to dry in the sun. If you are using an electric dehydrator, get a big bowl or bucket of ice water and dump some sugar or citric acid (vitamin C) in it. As you chop, put the finished slices into the bowl until you are ready to lay them on the drying trays.

3. Spread one layer on a drying tray. When making fruit leather, line the tray with plastic wrap or even parchment paper. Each type of food will dry at a different rate, so it's important to keep them all separate. Make sure you label everything so you know what it is.

4. Put the trays in the hot sun, or in the electric dryer. If you are drying meat in the sun rather than the dryer, it should be on a tray that is four feet above a slow fire. The firewood should be nonresinous hardwood and with very low flames because its purpose is to keep away the birds and the flies. Green wood that makes a lot of smoke works well for this.

5. Turn big chunks of food three times a day and small foods once or twice a day. In a dehydrator, move the trays of almost-dry food to the top and the

moist food to the bottom. It should not take more than two days for them to dry. Everything needs to be protected from dew and bugs. If you must continue the next day, bring everything inside before dusk and bring them out again in the morning when the sun is out.

6. Vegetables are dry when they are brittle and break when bent. Fruits are leathery or brittle, and should produce no moisture when squeezed. Fruit leathers will be a little sticky, but easily peeled from the paper or plastic lining in the tray. Pods will become dry and brittle. Meat should be hard and solid, with a uniform appearance; this means that they should be the same dark red color throughout, without any excessively large wrinkles in one place. Break a piece to make sure it is dry in the center.

7. Put fruits and vegetables into a wide-mouthed bowl for a week, stirring it two to three times a day. Keep it covered with a screen or porous cloth. This conditions the food to resist mold. Then repack it tightly in an airtight container or freezer bag and store in a dark, dry place.

8. If you want to pasteurize the food, put it in the oven for 30 minutes at 175°F. Remember to label with the type and date, and check it in the first two weeks for moisture—if you find that there is some, you'll need to dry them some more. Properly dried food should stay good for at least six months or more. If you find bugs, remove the bugs and roast the food at 300°F for 30 minutes.

Fermentation

Lacto-fermentation has become more popular again in recent years because it saves the nutritional properties of the food preserved and has all kinds of friendly bacteria. Where other types of food preservation techniques try to kill all the bacteria, fermentation encourages bacterial growth. It works because the fermentation process produces lactic acid, which kills botulism and other bacteria. For this reason, it can be much safer than canning or even eating raw vegetables, which can harbor *E. coli.*

It is highly recommended that you put the fermenting foods into jars with rubber-sealed lids. The rubber seals release gasses that build up during the fermentation process, preventing an explosion. Traditionally, people used crocks as well. You can sterilize the jars by pouring boiling water in them if it makes you feel better, but soap and water is enough. Grow the food yourself or get it from a farmer who has clean produce and a good reputation.

Pickling

Cucumbers are the most popular type of pickle, but beets, carrots, green beans, onions, radishes, Swiss chard ribs, turnips, zucchini, and many other vegetables can be pickled too. The recipe is pretty much the same as for kimchi and sauerkraut, with only a few variations. The recipe is as follows:

- 1 pound of sliced cucumbers
- 1 cup of sea salt for salting the layers
- Water to fill jar
- Salt water solution for soaking dill
- 1½ tablespoons of sea salt for brine
- 1 cup of unchlorinated water
- Mustard seeds
- Peppercorns
- Lots of fresh dill
- 2 cloves chopped or mashed garlic

Lay the cucumbers tightly into a bowl, adding the 1 cup of salt as you go, layer by layer so they are all salted. Fill with water so that there is at least an inch of water on the top. Soak the dill heads upside down in salt water as well. After 24 hours, mix the unchlorinated water and brine salt so the salt is completely dissolved. If they still have peels on them, you can poke holes in the cucumber peels with a fork. Pack the cucumbers tightly into extremely clean or sterilized jars by layering them with the mustard seeds, peppercorns, dill, and garlic. Add the brine solution, but don't fill it right to the brim. Optionally, you can place a horseradish leaf on the top to protect the top layer. Close the jar tightly and keep it in the kitchen for a couple of days so you can watch it. When it begins to form bubbles on the top, put it in the fridge or cellar for six weeks before eating.

Sugar

The first thing you need to make jam or jelly is pectin, which turns your juice into a thicker substance. Many people buy Certo Fruit Pectin in packets and add it to the *jam*, but you don't need to do that if you have apples.

To do this, save the peels and cores of apples and tie them up in a bundle of unbleached muslin cloth. The more apples you put in there, the more pectin you will have, but even a few will make a difference. There is also a distinction between jams and *jellies*. Jelly is the kind of spread that has no chunks of fruit in it. Instead, the fruit has been pureed and strained. Since you are using natural pectin, you have to make jam. The fruit in jam has simply been crushed or just boiled until it is mush. The measure of sugar has everything to do with the quality of your jam. Some fruits also

---

Does it have to be sugar?

Sugar as a preserving method is almost exclusively used for fruit, because of all the natural sugars that they already have. The extra cane sugar has the job of preserving the color of the fruit. It also helps to create the firm consistency we are familiar with in jams and jellies. It also helps the fruit last longer; without the sugar, the jam won't store as long. It doesn't have to be cane sugar, but cane sugar ensures success. A note about honey: it is recommended to substitute only up to half the sugar, or your jam will be a runny syrup that lasts a much shorter amount of time.

To substitute other sugars, use the following:

- honey: ¾ cup for every cup of sugar
- brown sugar: use the same amount as sugar, but it has a strong flavor—works well for peaches
- raw sugar: add ¼ cup more for every cup of sugar

---

need extra acidity to form a gel. Usually this is lemon juice, but any citrus could work.

Clean, peel, and remove the stems of the fruit you want to add. Grapes need a little extra processing to remove the peel, which can be done by simmering or squeezing the fruit. Rosehips are gathered in the winter, after the first frost, and must be pureed to remove all the seeds. Once they are all prepared, put all the ingredients in a big pot with the muslin bag of apple cores—with the exception of the rosehips, which don't need apple cores to thicken up. If you are making apple jam (or apple butter), you will have to add 1 tablespoon

---

Fruit	Sugar	Citrus
1 cup apples	¼ cup	1½ teaspoons
1 cup apricots	1 cup	2 tablespoons
1 cup berries	¾ cup	1½ teaspoons (optional)
1 cup grapes	1 cup	
1 cup peaches	¾ cup	2 tablespoons
1 cup pears	¾ cup	2 tablespoons
1 cup plums	¾ cup	
1 cup rosehips	¾ cup	

of water per cup of apples. Bring the fruit to a simmer over low to medium heat, stirring now and then. Toward the end you will have to stir more. Continue until it has reached a consistency that doesn't drip. Most jams, especially berry jams, may still be fairly liquid and won't have that store-bought jiggle, but as long as it slides slowly off a spoon it should be done.

Fill the jars up as full as you can with the hot jam, close tightly, and turn the jars upside down. The jam will help sterilize the empty space at the top. Store them upside down and they will last the winter. Other ingredients to add include cinnamon, fresh walnuts, raisins, mint, currants, hazelnuts, and vanilla.

Good ingredients for jam:
apple
apricot
blackberry and raspberry
blueberry and cranberry
cherry
citrus
grape
peach
pear
quince
strawberry

Apple jam:

1. To use the natural pectin in the fruit, first clean the fruit and taste it for tartness. If it is not very tart, add some lemon juice for extra acid. Prepare the apples as you would to make pectin. Cut up enough apples to make 5 pounds, add 1 cup of water, and simmer (you may have to add more or less water for a better simmer).

2. Stir frequently until soft. Sieve the pulp in cheesecloth or a Foley food mill.

3. In your jelly pot, measure in 10 cups of pulp, ½ cup of water and 2½ cups of sugar.

4. Cook the pulp quickly over medium-high heat (220°F), stirring constantly. When it is done it should be boiling fast, have a sheen, and it will have reduced in volume. This is tricky . . . if you stop the boiling too soon it will not set, but too late it will be rubbery. You can test to see if the jam is ready to set by removing a small amount of it and putting it on a plate. Stick it in the freezer for a few minutes. If it gels, it is ready.

5. Sterilize the jars as you would for canning, then put the jam in hot and seal them. The recommended processing time is 5 minutes at sea

The Ultimate Guide to Urban Farming

level, 10 minutes at 1,000 feet, and 15 minutes above 6,000 feet (see the section later on canning). This can make 11 cups of jam.

Chutney formula:

Chutney originated in India and is usually prepared right before a meal, but it can also be used to preserve any kind of fruit or vegetable. It is very similar to jam but there are a bunch of different spices and flavors combined together, and rather than being something reserved for toast, chutney is used to top cold meat, potatoes, rice, and salads.

You will need the following:
- 4 cups of chopped fruit or vegetables (apples, mangoes, plums, tomatoes, rhubarb, etc.)
- 1 cup of a complementary vegetable (radish, zucchini, eggplant, etc.)
- 1 cup of chopped onions
- 2 tablespoons of brown sugar
- ½ cup vinegar
- Salt
- Herbs (ginger, mustard seed, cloves, cayenne pepper, rosemary, pepper, curry etc.)

Throw all the fruits and vegetables and onions into a pot with the herbs and salt to taste, add a little water, and bring it to a boil. Simmer until everything is very soft and mixed together. Add the sugar and vinegar and continue boiling until it has the same consistency as jam. Sterilize the jars and lids, and pour the chutney in while it is very hot. Close the lids immediately and store upside down. Ketchup is basically a chutney.

Vegetable Oil

Any oil extracted from a plant is a vegetable oil. This includes almond oil, avocado oil, castor oil, coconut oil, hazelnut oil, olive oil, wheat germ oil, soy oil, sunflower oil, canola oil, peanut oil, sesame oil, etc.—any nut, grain, bean, seed, or olive. The oil is pure fat. Only sesame seeds and olives can be pressed without being heated. This is called "cold pressing" and is the most nutritious. Other foods need to be heated before pressing. There are several modern ways to extract the oil by pressing, either hydraulically or by expeller. The other way is with a solvent, which is definitely bad for your health. Most store-bought oil has been refined to the point that its age and taste are gone. Homemade oils are rich and flavorful, and it is easy to tell when they have gone rancid (store-bought and processed oils are more difficult to tell). When you purchase olive oil at the store it is labeled Olive Oil, Pure Olive Oil, Virgin or Extra-Virgin. Virgin means that no chemicals were used in extracting the oil. Regular olive oil has some virgin oil mixed in, but isn't very high on flavor and has had solvents in it. Extra-virgin has exceptional taste.

With olives it is just a matter of crushing the juice out and letting the oil rise to the top. With other foods, first crush them and press the oil out, then boil the pulp in water and more oil will rise to the surface and can be skimmed off. There are no home-size presses for this, so either a fruit press can be used or a homemade press can be devised for this purpose.

Vinegar

Vinegar is a useful ingredient for many dishes. It also preserves foods like herbs and is a great product to sell. The easiest way to collect vinegar is to keep a mixture of half vinegar and half cider at 80°F for a few days. The thin scum on the top is "mother of vinegar." You can save this to turn any fruit juice into vinegar.

1. Take sweet apple cider, uncooked with no preservatives, and fill a gallon glass jug to the neck. The jug should have an airlock with a tube. As the juice ferments, the carbon dioxide will come out the tube and bubble the water.
2. Keep the cider at room temperature and then wait 4–6 weeks for it to ferment. When the bubbling stops, pour half the cider into another jug. Then add mother of vinegar to each by putting a little on a dry corncob and floating it inside, or add already-made vinegar to it (1 part vinegar to 4 parts cider). Cover each jug by tying a cloth on over it.
3. Try to keep the jugs at 70–80°F. Ordinarily it will take 3–9 months. When it is vinegar, dilute it before you use it.
4. To store it, strain the vinegar through cheesecloth and store in a cool, dry place in bottles.

Test your vinegar to make sure it is strong enough (titration):

1. Mix a small amount of baking soda in water in a small jar. There should be enough soda that some of it settles to the bottom.
2. Steam a head of cabbage in a little water and keep the juice (make sure it is very purple liquid). Pour this juice in another jar.
3. Add a few ounces of water to two drinking glasses, making sure they're equal to each other. Using an eyedropper, add enough cabbage juice to each of the glasses to make them purple (put the same amount in each).
4. Rinse the dropper, and then put seven drops of five-grain store-bought vinegar into one of the glasses of purple water.
5. Rinse the dropper and add seven drops of the homemade vinegar to the other glass.
6. Rinse the dropper, and then add 20 drops of baking soda water to the store-bought vinegar mixture. This will turn the water blue.
7. Add baking soda one drop at a time to the homemade vinegar mixture, counting each drop, until it is the same color blue as the store-bought one.
8. To calculate the acidity, divide the number of drops you used by four. Thirty drops divided by 4 = 7.5 percent acidity.

Your vinegar is a little strong to use in any recipes at this point. The following formula uses 7.5 as the example homemade vinegar acidity. Most recipes use 5 percent acidity, so that's probably what you'll dilute it to.

1. Subtract 5 from 7.5 (your current acidity) to find the difference: $7.5 - 5 = 2.5$
2. Multiply the answer by your total amount of vinegar (in ounces). 1 quart = 32 oz. $32 \times 2.5 = 80$
3. Divide the answer by 5. $80 / 5 = 16$
4. The answer is how much water you add to the vinegar. So in this case, you would add 16 ounces of water to it.

Now you can pasteurize your vinegar. Pasteurizing prevents you from using the vinegar as a starter, but it also keeps it from being cloudy, because no mother forms. Put the bottles (loosely corked or unsealed) into a pan filled with cold water. Heat the water gradually until the vinegar is about 145°F and keep it there for 30 minutes, then cool.

Herbal vinegars are the most fantastic way to use and sell your homemade vinegar. Collect some aromatic herbs. You can use the leaves, stalks, flowers, fruits, roots and even nuts. Cut them up very finely, and fill any jar up to a thumb's width from the top. Pour room temperature vinegar into the jar up to the top—apple cider vinegar is best. Cover it with a plastic lid, several layers of plastic wrap or wax paper held on with a rubber band, or a cork. Don't use metal, as the vinegar will corrode it. Label it with the name and date, and in six weeks it will be ready to use. The fun part about this is that not only is the vinegar useful, it can make a great gift because the jar can be beautiful. To use, pour on beans and grains as a condiment, in salad dressing, in cooked greens, stir fry, soup, or any recipe that calls for vinegar. You could even use it for cleaning. After six weeks you can decant the vinegar into a better jar so it does not get stronger.

Best tasting herbs:

Apple mint (*Mentha sp.*) leaves, stalks
Basil (*Ocimum basilicum*) leaves
Bee balm (*Monarda didyma*) flowers, leaves, stalks
Bergamot (*Monarda sp.*) flowers, leaves, stalks
Burdock (*Arctium lappa*) roots
Catnip (*Nepeta cataria*) leaves, stalks
Chicory (*Cichorium intybus*) leaves, roots
Chives and especially chive blossoms
Dandelion (*Taraxacum off.*) flower buds, leaves, roots
Dill (*Anethum graveolens*) herb, seeds
Fennel (*Foeniculum vulgare*) herb, seeds
Garlic (*Allium sativum*) bulbs, greens, flowers
Garlic mustard (*Alliaria officinalis*) leaves and roots
Goldenrod (*Solidago sp.*) flowers
Ginger (*Zingiber off.*) and wild ginger (*Asarum canadensis*) roots
Lavender (*Lavendula sp.*) flowers, leaves
Mugwort (*Artemisia vulgaris*) new growth leaves and roots
Orange mint (*Mentha sp.*) leaves, stalks
Orange peel, organic only
Peppermint (*Mentha piperata* etc.) leaves, stalks
Perilla (Shiso) (*Agastache*) leaves, stalks
Rosemary (*Rosmarinus off.*) leaves, stalks
Spearmint (*Mentha spicata*) leaves, stalks
Thyme (*Thymus sp.*) leaves, stalks
White pine (*Pinus strobus*) needles
Yarrow (*Achilllea millifolium*) flowers and leaves

Canning

Note: If canning isn't done properly, you can end up with botulism and die. Follow the instructions exactly, and use proper jars and seals.

The initial investment in canning equipment can be steep but worth it. A pressure canner has a dial-type temperature gauge, pressure regulator, and lock-down clamps. It is not recommended to can without an accurate pressure regulator. In the old days it was done, but today, and for beginners, always use one. This is because water boils at 212°F, not hot enough to kill all bacteria, so pressure is used to raise the temperature to 250°F.

An enameled canner is blue or black with white speckles, has a lid and canning rack, and is for sterilizing and boiling things. It is also used to can things such as fruit that has lots of acid so it's safe to can—which is called water bath canning.

The jars must be heavy mason jars (or similar) made for canning. Using old mayonnaise jars doesn't work because they break very easily under pressure (although it is possible in an enameled canner). Some companies are now selling classic mason jars with the wire bail lids, but the cheapest method is to get reusable jars and get new lids and screw bands. This type has a small lid with a rubber seal, and a metal band or ring that tightens it (the rubber seal can only be used once, but the ring can be reused). Then you throw out just the lid after one use. Never use cracked, chipped jars or jars with a worn rim.

Pressure canning:

1. Prepare the food for canning by blanching, skinning, pitting, slicing, and poaching as needed (use a canning recipe for each mixture). Use only the

---

Equipment needed for canning:
  pressure canner
  enameled canner
  glass canning jars
  a sieve
  canning jar lids and seals
  wide-mouthed funnel
  rubber gloves
  jar tongs or lifters
  a loud timer

---

best fruit, and make sure to remove all the bruised or infected parts.

2. Boil all your plastic and stainless steel equipment for 30 minutes, then wrap in a clean towel. Wipe down counters with a chlorine scouring powder (like Comet) or other antibacterial solution, and rinse with boiling water. Dip knives with wooden handles in boiling water. Wash canning jars and put in simmering water (180°F) for ten minutes. Heat canning lids in hot water but don't boil.

3. Put the food into the canning jar and fill liquid up to a half inch from the top (for air space). Put on a lid snugly, but not so tight that air can't escape.

4. Put the jars in a rack in the pressure canner with 2–3 inches of boiling water in the bottom. Make sure the jars don't touch the sides, the bottom of the canner, or other jars. Fasten the lid and open the petcock. For fruit: Put jars in the rack, and cover with 2 inches of briskly boiling water. Put the lid on, but don't fasten it down. Leave the petcock open so that steam can escape.

5. Turn on the heat until steam comes out of the petcock in a steady stream (about 10 minutes). When the steam is

---

nearly invisible 1 or 2 inches from the petcock, close the petcock.

6. Raise pressure rapidly to 2 pounds less than you need, then lower the heat and slowly bring the pressure up the remaining 2 pounds. This slow pressure rise should last for the amount of time specified for that particular food. Turn off the heat and let the pressure drop to zero. Wait 2 minutes, and then slowly open the petcock. Open the lid away from you so you won't get burned by steam, or just wait a while until it cools.

7. Pull the jars out with tongs, holding them straight upright (don't tip them!); let them cool on a dry, nonmetal surface for at least 20 hours. When they are cool, test the seals; wash, dry, and label them; remove the jar rings; and store in a cool, dry, and dark place.

Pressure Canning Altitude Adjustments

Find out your altitude, and for every thousand feet use the appropriate pounds. The poundage raises the temperature (which takes longer the higher you are).

Altitude in Feet	Weighted Gauge	Dial Gauge
Under 1,000	10 pounds	11 pounds
1,000 to 2,000	15 pounds	11 pounds
2,000 to 4,000	15 pounds	12 pounds
4,000 to 6,000	15 pounds	13 pounds
6,000 to 8,000	15 pounds	14 pounds
8,000 or more	15 pounds	15 pounds

Water bath canning:

1. Use only highly acidic foods such as high-acid tomatoes and tomato sauce (no mushrooms or meat), jam, jelly, juices, barbecue sauce, chili sauce, relish, pickles, etc., or use a recipe that calls for an acidic additive specifically for water bath canning in an enamel canner. If you are in doubt of the acidity of something, add 2 tablespoons lemon juice or ½ teaspoon citric acid (vitamin C) per quart.

2. Clean and sterilize the jars by boiling and heat lids (same as step 2 for pressure canning), then fill the jars with the hot food (as called for in the recipe). Put a new lid on the jar and screw on a ring firmly.

3. Fill the canner with water. The water will reach 1–2 inches above the top of the tallest jars. Bring to boil, and put the jars into the wire rack in the canner. This is the water bath. Put the cover on and let it reach a full rolling boil.

4. Processing time starts when the water reaches full boil. At the end of the processing time (check the recipe), lift out each jar carefully and place on dry folded towels. Each jar will seal and you will usually hear a ping as the lid is sucked in. Don't touch them until they are cool.

5. The next day (usually takes a long time to cool), remove the rings, check for a good seal, wash the jar, label it with contents and date, and put it in a cool, dark, dry place.

Water Bath Canning Altitude Adjustments

Altitude in Feet	Increase Processing Time
1,000 to 3,000	5 minutes
3,000 to 6,000	10 minutes
6,000 to 8,000	15 minutes
8,000 to 10,000	20 minutes

Food safety rules:

- Never eat from a jar that has lost its seal. It won't make a suction noise when you open it, and the lid won't be sucked in.
- When canning, make sure the temperature is high enough and that you boil it for a long enough time.
- Keep the jars stored below 40°F.
- Cook canned food for at least 10 minutes at boiling or 350°F in the oven before eating.
- Don't use recipes or methods from before the mid-1980s—always use modern recipes.
- Throw out anything with mold—don't try to scrape it off. Store jars loosely to prevent mold.
- Don't use jars larger than a quart because they can't be heated enough.
- Check the top rim of the jars for nicks before you use them.
- If your boil drops or your temperature drops at any time, start over completely from the beginning.
- A boil means a super-hot, really bubbling boil—not tiny bubbles.
- Put hot food in hot jars and cold food in cold jars. If not, temperature changes will break the glass.
- Don't put hot jars on a cold surface or in cold air.
- Check the processing time for altitude and adjust as needed.
- If your water bath is not covering the jars by at least an inch, cover with more boiling water.
- Add a piece of tomato to everything you can. Tomato has enough acid in it to prevent botulism from occurring.

Government authorities say that canned food lasts one year, and that's when you should throw it out. However, most home canners use their food far after that time period without problems. If you follow all the safety rules, your food should be fine for a very long time, even ten to twenty years. However, it will not have the same nutritional value, and the longer you wait, the bigger the risk.

Live Storage

Pumpkins, potatoes, dry beans and peas, onions, parsnips, turnips, apples, oranges, pears, tomatoes, and most other root vegetables can be stored *live*, or without any processing, in a properly maintained cold storage. While the conditions of your cold storage are essential to your success in this, picking a species made for preserving this way and harvesting at the right time are extremely important. Make sure that the foods you use are not bruised or blemished in any way, and remove the tops.

Food	How Long	Method
Apples (especially Winesap Granny Smith, Black Arkansas, Idared, Liberty)	4 months	In small crates stacked no more than 2–3 high. Stack the ripest ones on top. Store above ground on a shelf or table.
Cabbage	3 months	Pick before frost, removing roots and outer leaves. Place upsidedown in a single, loose layer in a crate. Stack the boxes and cover with a tarp.
Carrots	4 months	Line crate with leaves and stack carrots upright against each other. Stack crates above ground.
Chestnuts	6 months	Soak nuts in water for 2 days. Remove anything that floats. Let dry for 1 day on a screen out of the sun. Store in bucket with sand with a screen cover.
Leeks	Varies	Cut off roots and leaves. Transplant to a container of sand or sawdust and water once during winter.
Root Vegetables	Varies	Root vegetables need to be put into any large waterproof container and layered with sand or sawdust so they are not touching each other.
Squash	3–8 months	Wipe down with vegetable oil and wrap loosely in newspaper. Throw out moldy ones.
Tomato	4 months	Pull up the entire plant at the start of autumn. Wrap each tomato in newspaper and hang the plant upside down. The green ones will ripen.

The first step is to leave the dirt on, which will help protect them from decay. Use plastic buckets, or enamel cans, as these will be rodent and decay-proof. Fruits need to be stored away from vegetables because the gas produced by apples can cause vegetables to sprout. Pack root vegetables in damp sawdust, sand, or moss. Keep potatoes out of any light or they will turn green and become poisonous to humans. The table on page 146 shows how to store a variety of foods and how long they will last in live storage.

Put a thermometer on the inside and outside of the cellar and monitor the temperature every day. Use doors and windows to maintain a temperature of 32°F: open the door in cold weather and close the door in very cold or hot weather. Alternatively, you could install a fan attached to a thermostat that functions similarly to the kind used to ventilate a greenhouse. The food needs humidity so that it doesn't dry out, usually 60–75 percent. If necessary, you can set out pans of water, sprinkle the floor with water, or cover the floor with damp sawdust. If it is too damp, take pumpkins, squash, and onions to a dryer area or they will rot quickly. Remove all spoiled food, and if something is about to spoil, dry it quickly before it rots. If something is rotting or molding, get rid of it immediately, and make sure there are no insects infesting anything.

If you can't build a cellar or cold storage, you can use a *clamp*. This is a very old device that functions similar to a cellar and just leaving roots in the ground. It can range from just a hole in the ground to an old washing machine drum to a nice brick box sunk into the ground. The pit can be between eight and twenty inches (20–50 cm) deep and lined with something to stop rodents. This can be wire mesh, clay, or brick. Inside, throw down a layer of sand, leaves, straw or twigs, and begin layering the vegetables with layers of dry material in-between. If you don't have very many to store, just put them all in one pit (not packed too tightly), but if you have lots, make more than one clamp, one for each type of food. In the center, leave a hole or tunnel up to the top and fill it with twigs for ventilation. Cover the top with another layer of dry material and cover that with

Beet root	Leave in the garden until very cold.
Brussels sprouts	Cover well with dry straw and a sheet of plastic.
Cabbage	Dig an 8-x8-inch trench running east to west. Lay down a row of cabbages in the trench with the stem toward the south. Cover with straw.
Carrot	Cover well with dry straw and a sheet of plastic.
Cauliflower	Leave in the garden until very cold.
Chicory	Cover well with dry straw and a sheet of plastic. Prevent rot by uncovering in mild weather.
Curly kale	Cover well with dry straw and a sheet of plastic.
Endive	Cut off the leaves, cover with 8 inches of dirt. Cover shoots with more dirt. Eat in early spring.
Jerusalem artichoke	Leave in the ground and cover with straw.
Kohlrabi	Leave in the garden.
Leek	Cover well with dry straw and a sheet of plastic.
Lettuce	Dig a 16x16-inch trench and lay the heads in but they should be not touching. Cover with straw.
Parsnip	Leave in the ground and cover with straw.
Radish	Cover well with dry straw and a sheet of plastic.
Salsify	Leave in the ground and cover with straw.
Turnip	Leave in the ground and cover with straw.

The Ultimate Guide to Urban Farming

a wooden board. Cover that with plastic and put a heavy rock over the top to keep animals out.

Some root vegetables can simply be left in the ground. This is similar to a clamp except that they are just individually surrounded by the soil they grew in. They must be protected from frost, however, and this is done in different ways depending on the vegetable. You can do this in October or November, before the first frost. If you have raised wooden beds then you simply need to cover the plants in the manner described below; if not, you will need to sink wooden boards into the ground around the bed to help protect it. The table on page 146 shows how you can leave certain vegetables directly in the ground until you are ready to eat or sell them.

Saving Honey

Honey does not need to be frozen, canned, or refrigerated, and it keeps well in any type of container. Don't refrigerate it or it will crystallize sooner. When storing for a long time, don't let it get warmer than 75°F or it will lose flavor. Use a container that has a wide mouth because eventually the honey will crystallize and then you won't have to be pouring it, you can scoop. When you keep a small amount in your kitchen keep it in a warm place.

Freezing. There is no real need to freeze honey. Freezing will prevent honey from crystallizing if that's necessary for you, but there is a solution to crystallizing as well. Just warm it up and it will liquefy.

Crystallizing. This happens naturally to honey, and it simply dries. It can be used exactly the same in a recipe, and to make it liquid you simply warm it up to 130°F as quickly as you can and then cool it as quickly you can. If it is in a can, put it on your woodstove. If in a jar, put it in a double boiler. Don't let it get hotter than 130°F.

SAVING SEED

"Passion is of the nature of seed, and finds nourishment within, tending to a predominance which determines all currents towards itself, and makes the whole life its tributary."
—*George Eliot,* Daniel Deronda

Each tiny seed carries thousands of years of evolutionary history, waiting until just the right moment to burst into life. Within a few days, a dormant little speck no bigger than a grain of sand becomes a white sprout stretching up toward the light, and only weeks later is a tall green ready to be eaten. Each variety carries the potential of continuing the species despite predators, weather, and disease, and each species could be the possible salvation of some human society facing starvation.

A Little Science

Just like our interaction with the soil, interacting with plants requires a little scientific knowledge. To make seeds, plants must *flower.* It takes male and female flower parts to make a seed, and for some species these are on the same plant. Others have male and female parts on separate plants. The pollen from the male part must make it to the female part and thus seeds are born.

For serious seed savers, your first concern is to make sure your pollen isn't contaminated by another variety. This is the big problem with GM genetic drift. The pollen containing all of that genetic material mixes up the DNA of your own strain, creating something else entirely. You'll have to have a basic understanding of how your particular plant gets pollinated and how to keep it isolated. The wind and bees are the biggest culprits in contamination. Only grow one variety of the species you want to save. Bear in mind the distance between you and other gardens. Is your neighbor growing carrots? Each variety has an optimum isolation distance, which is information you can find out from a quick Internet search. If this is not possible, plant earlier or later than your neighbor so that your plants flower at different times.

You can only save seeds from open-pollinated varieties, which is usually indicated on the packet if you purchase it from a good source. Hybrids (or F1 hybrids) will produce seeds that are not true to type—that is, the child of a plant grown from a hybrid seed may not look anything like its parent. This is because they are like clones and genetically identical, and if they have children they will all be inbred and look like the grandparent plants. Open-pollinated varieties will carry the traits of the parent plants, although there may be a lot of variation. There's nothing wrong with hybrids, it's just difficult to breed them.

You can isolate the flowers of a plant by covering them in bags made out of mesh or paper, but this is not the most ideal method. The bags can get in the way once you start pollinating and can increase the chances of mold or wilt or other problems. The simplest way to ensure pollination is to do so by hand, literally. Use your finger or a paintbrush and mimic the actions of a bee. It may seem like a lot of trouble, but each plant will produce thousands of seeds and careful pollination ensures that your plants stay pure.

Once you are well on your way to serious seed saving, you'll learn how to "rogue out." Rogueing out just means saving only the seeds from plants that are true to type, ones that carry the traits of the parents. This takes some practice but is the most important part of saving good seed. If a carrot variety grows long, has a bright white root, and is resistant to carrot fly, then you would certainly not save seed from a short, orange carrot that was injured by carrot flies. Sometimes the discrepancies are more subtle . . . you decide not to save the seed of the tomato that was less red, or even less sweet.

When your true-to-type plant successfully makes seeds that have been appropriately isolated from contamination from other varieties, it's time to save them. Gather the seeds, and lay them out to dry. For some seeds, you'll have to first remove any pulp and wash them off. Although some people use a dehydrator, heat is a bad thing because high temperatures can either kill the seed or make it germinate early. It works just as well to lay them out on a surface that is low in humidity. A large board, screen, or pan is ideal. Paper and cloth will just stick to the seeds, so just keep the surface bare. A gentle fan to keep air circulating in the room (but not blow directly on the seeds) can help speed the process, which generally takes a couple of days to a week.

Once dry, the seeds should last about a year, if stored properly. Put them in a clean, dry container. It does not have to be airtight. In fact, sealing too tightly can increase the chances of mold. If you plan to use the seeds within a few months, store them in a cool, dry, and dark location. If you need to store them longer than that, put them in the refrigerator.

Plants that you can save the seeds from:

Don't buy hybrid seeds or patented varieties (they should specify on the package). Heirloom seeds work great, but for normal varieties buy the open-pollinated kind. Open-pollinated seeds were pollinated naturally, and not in a lab. When you have diverse seeds that were grown naturally, they are more resistant to pests and disease. Hybrids get bigger but die more easily. Some plants, such as bananas, potatoes, and some herbs don't need seeds to grow; you simply stick a piece of the plant into the ground (a process called propagation).

Saving seeds:

1. First make sure your plants get pollinated. This means either encouraging bees, or sticking your finger in one flower and rubbing the pollen in another.
2. Flowers that make a flower head will create a seedpod. Simply cut it off, dry it, and break it open when you need to plant. Fruit are another example of a pod, except you can eat them. Cut

---

Types of plants:

Annuals: Makes a flower and then makes seed for each flower in one year.

Biennials: Most root vegetables are biennial. They flower and make seed after two years. The first year they just store up food in a root.

Self-seeders: Self-seeding plants create volunteers, or plants that grow without you having to do anything; for example, a sunflower drops its seeds before you get to them.

---

open the fully ripe fruit and scrape out the seeds, and let them dry. For any other plant, make sure there are no weed seeds in with your good seeds by making a sieve. The mesh should only allow the seeds you want to keep to stay on top.

3. After putting them through a mesh and sifting, dry the seeds for at least two weeks before storing them, unless the seeds have beards or tufts.

Let them dry well, and then store them in a cool, dry place such as a dry cellar or a basement. The fridge is good if you keep them in tightly sealed, very dry jars and containers.

4 | Plants

▲ Radishes aren't very valuable on their own, but they add a ton of value to a salad mix and look really great in a market stand.

CASH CROPS

"Commercial agriculture can survive within pluralistic American society, as we know it—if the farm is rebuilt on some of the values with which it is popularly associated: conservation, independence, self-reliance, family, and community. To sustain itself, commercial agriculture will have to reorganize its social and economic structure as well as its technological base and production methods in a way that reinforces these values."

—*Marty Strange,* Family Farming: A New Economic Vision

Berries

While berries grow in almost every climate, they are most suited to cool and temperate zones, where they grow everywhere naturally. Cranberries, blueberries, raspberries, and strawberries are all pioneers, but unlike most pioneer species, they also offer a valuable edible product. They improve the soil and shelter seedlings from invading deer. They also grow in soil that isn't very good and need little care except frequent picking. They should be provided with liquid manure and thick mulch to keep the grass from growing. They need to be protected from birds, which will quickly eat 30 percent of your cash crop. Berries must be picked every day, so if you are within reasonable driving distance of a city, you can offer them as a *U-pick* product.

Bird protection can be done with a mesh cage, which will also house a polyculture system. The raspberries or boysenberries grow on trellises to save space, and blueberries can be intercropped on raised beds two feet (0.6 m) high and five feet (1.5 m) wide with drainage at the base, preferably with piping. Strawberries are grown as a groundcover. Lizards, frogs, and quail are released into the cage to control insects (and the quail can provide food). This type of system must be somewhat small, but because there is very little crop loss, it may be more profitable.

If you decide you want to grow a bigger cash crop and go the U-pick route, bird-deterrent kites that look like hawks are available and are simply tethered around the field. Their effectiveness is questionable, but if they are removed after the harvest the birds will hopefully not get used to them. The advantage of having a bigger field is that people will do the work for you, but you must build wider, grassy paths between the three-feet-high (1 m) raised beds. You will also need buckets, scales, and bags for packaging.

Both of these systems use drip lines for irrigation. It is quite common in North America for berries to be watered with sprinklers, which are much easier to install and possibly cheaper, but during the summer you will be watering

▲ Honeyberries grow great in containers with proper drainage and will produce a lot of fruit if you have several varieties.

twice a day. Sprinklers must be left on for hours, letting much of the water evaporate in the sun, and some of it won't even make its way down to the soil. Drip line uses less water and conserves it by depositing it directly near the roots.

Blackberries are tasty but also very likely to get out of control. It is probable that there are some blackberries on your property already, which you may want to utilize as a hedge or barrier somewhere, but you will have to form a strategy for keeping them under control. Removing them takes years so it's a good idea to start right away. In a very small area it is possible to cut them back and cover them with a strong mulch of tough plastic weighed down with rocks or other material. Once they have rotted there for two years, you can dig the roots up. An area of blackberries of a quarter acre or more can be fenced off and used as a pig field at a ratio of twenty pigs per acre (0.4 ha). The next year twelve goats per acre (0.4 ha) are grazed there, followed by pigs again the next. The blackberries won't return as long as you keep something in there like sheep or goats, or plant trees or a hay cash crop. Another strategy is to plant apples, figs, pear, and plum trees in the middle of the blackberries forty feet (12 m) from the edge. In about five years these trees will have grown enough that you can let cattle in to graze. The cattle will eat the windfall fruit and trample down all the blackberries.

Cultivating Wild Edible Foods

Obviously it would be devastating (and any wild food book will tell you this) if everyone ran into the woods and started grabbing up wild plants, especially since many of these are endangered. But wild plants are full of nutritional value and, once you identify what grows in your climate, can be simple to grow. Starting out cultivating native species can be tricky especially if you have a landscaped yard. You'll have to rip out any nonnative plants that will compete and carefully choose what are going to put in to make sure it is appropriate for your soil and climate. It's also going to be a challenge creating a small ecosystem, especially if you want to grow fish and water plants.

Getting the seeds and plants initially can also be a challenge. These need to be locally sourced plants and seeds, but not pulled directly from the wild. Your best bet is to search Google for "native plants" or "wild heritage plants" in your province or state. You could also ask your local nursery or local county extension. Make a list of the plants you want, and then search for those specific species, because most places will carry mainly ornamental plants, with one or two edible varieties. But once you have

your first plants, most of these wild varieties simply spread on their own.

Fish and Water Plants

Building a backyard pond is a common practice, and you can raise fish in one. Many people raise channel fish, trout, striped bass, and tilapia. Some people raise a couple of these, although most choose channel fish and tilapia. For a pond that feeds itself, tilapia are the easiest because they eat mostly plant material and plankton. You can introduce small, edible, cleaning plants to your pond, such as duckweed, which will feed the fish and clean the waste at the same time. Some are experimenting with cleaning gray water with duckweed, which could be worked into the design as well. If you were to throw a decent number of breeding tilapia into your pond, they would reproduce and choke the pond as they overpopulated, so regular fishing or trapping to keep a balance is the key. Extras can be dried or smoked.

Mushrooms

Some people like to spontaneously try wild mushrooms, but they should only be eaten if picked by an expert. It's actually quite simple to grow mushrooms, because all mushrooms are wild mushrooms, and as you can see when you walk in the woods, they grow everywhere. The key to growing them outside in your little ecosystem is to replicate what they grow on in the wild, and then buy the fungus plugs from a mushroom supplier so that you know what you are getting. For example, in the Pacific Northwest (and many other places), chicken of the woods is a mushroom that likes to grow mostly on oak, though it also likes yew, cherrywood, sweet chestnut, and

▲ This mushroom farm uses stacked bags to hold their growing medium.

willow. Get a log of that wood, preferably an old cheap log that is rotting a bit. Soak it for 48 hours, and drill holes all over, which you then put the plugs into. Then you seal them with melted cheese wax, and bury the log about one-third into the ground. Make sure the log stays moist, and by fall you should have mushrooms. Chicken of the woods is particularly valuable because it can be prepared like chicken and contains very high protein.

Growing mushrooms indoors is a more common practice, especially shiitake, oyster, and white button mushrooms. The spawn or spore is purchased from a good supplier, and the growing medium depends on the variety you choose. Shiitake need hardwood or hardwood sawdust, oysters need straw, and white button usually need manure. The process is the same for all mushrooms: put the flat of *growing medium* (the material the plant grows in, such as coffee grounds) in a dark, cool place, such as a basement or at least a closet. Raise the

temperature to 70°F and plant your spores. In three weeks the spores will have rooted. Then decrease the temperature to 55–60°F, and cover the spores with about an inch of potting soil. Cover that with a damp cloth. You will need to keep the cloth and soil damp, so spray it down when it dries out. You will have baby mushrooms in around three to four weeks, and they will be ready for harvest when the cap is open and has fully separated from the stem.

You can also grow oyster mushrooms in coffee grounds, so they tend to be a good starter project. Normally, you would have to sterilize the straw, but coffee is already sterilized. There are excellent kits available that offer a growing bag, which holds the coffee grounds and is specially designed just for this purpose.

If you see any green during the growing process, you have a bad mold. If it's just a little spot, put salt on it to see if you can get rid of it, but if it all turns green, throw it all out.

The Ultimate Guide to Urban Farming

GRAINS

The Rice Paddy

It has been emphasized throughout this book that turning the soil should really be avoided as much as possible. Grain is a staple of most people's diets, and yet how do we grow it without destroying the soil? Modern grain farming uses monstrous machines to turn and fertilize the earth, and even bigger machines to harvest perfect rows of grain, and then the land is left to sit over the winter. This inefficiency can be easily solved with Masanobu Fukuoka's no-till method. Many sustainable farms follow a *rotational* planting schedule that involves planting legumes before and after the grain crop and includes a *fallow* period, a time where nothing is grown, to let the soil rest. The Fukuoka system, on the other hand, grows grain and legumes together continuously.

Not only is this system very sustainable, it requires very little energy. There is no mechanization and the energy used in human labor is also very low. The farmer can eat an average 2,000 calorie-per-day diet and still produce 1,300 pounds (590 kg) of rice (or 22 bushels) on a quarter acre. Using animals for labor in the traditional manner uses at least five times as many calories with the same results. Using a tractor uses at least ten times as much.

The other benefit of this system is the small space required. Many self-sufficient homesteads and small farms avoid growing grains because they have seen large commercial grain fields and assume that it must take vast expanses of land to get any amount of grain. So they grow potatoes and keep a cow instead. But in reality, to keep one human alive entirely on one food would take the following:

- 1,800 square feet of just grain
- 5,400 square feet on potatoes alone
- 13,500 square feet of just dairy farming
- 36,000 square feet of pigs alone
- 90,000 square feet of just beef

▲ A rice paddy with a low retaining wall that also acts as a pathway.

This is the greatest argument for a plant-based diet. For the amount of calories we need to live, we grow grain to feed beef, which is the least efficient thing we could do.

On the outer perimeter of your grain fields, a band of weed-control plants should be grown, such as comfrey, lemongrass, or citrus. These should be mulched with sawdust for even more protection. The grain strategy outlined here centers around rice and the construction of a paddy. If you can't have a paddy that fills with water, dry rice species exist that just need to be watered and have the additional benefit of being able to survive on monsoon rains alone.

First, level the ground and build a low mud retaining wall around the plot that can hold two inches (5 cm) of water. You may need to use a chisel plow the first year if the soil is extremely compacted. Spread lime or dolomite and a thin layer of chicken manure or compost over the area and water it. This soil disturbance and fertilizing only needs to be done once. Grain doesn't like rich soil so don't be overzealous; if it is too rich, the grain gets too tall and falls over (called lodging or going down). Don't walk in your field at all after the grass comes up, even to weed. To prevent weeds, plant buckwheat or amaranth (they grow fast),

or cultivate a lot before planting, or plant a lot of grain to offset losses. To water, wait till it rains, use flood irrigation, or set up a sprinkler beforehand.

No-till grain could also be called the no-work method. The four principles of natural farming are no cultivation, no fertilizer, no weeding, and no pesticides. It is important to understand here that we are not talking about soil that is never touched, but rather soil that is aerated and loosened by natural means only. With this mindset, we have to question weeding as well. What are the weeds doing for the soil? If the plant you want to grow is not being harmed in any way by the weed, then why pull it?

1. A variety of plants can be grown together. You may want to have several plots with different combinations, but each plot will always grow rice and white clover. Then you can add rye, barley, millet, winter wheat, or oats.

2. Rice seed is sown in early fall. It can be broadcast and covered with straw, or made into seed balls. Seed balls can be made in two ways: mix the seeds with mud and press them through a wire mesh, or you can wet the seeds down and roll them in fine clay dust until they form a ball shape. This clay mixture can be made from a combination of potter's clay, compost, and sometimes paper mush.

3. In mid-fall, harvest last year's rice and lay it out on rice racks to dry for a couple of weeks. Thresh off the husks and straw and save them.

Cold Climates

In some areas, it is just too cold to grow rice and you will have to use a system with shorter cycles. Spring wheat can be planted in the spring with oats, barley, or wheat as the winter crop. You can also experiment with squash, melons, tomato, cotton, vetch, or sunflowers as a no-till crop.

Type	Seed quantity per ¼ acre
Clover	¼ pound
Grains	1½–3½ pounds
Rice	1¼–2½ pounds

4. Within a month of the rice harvest, sow unhusked rice in the field, and spread the husks and straw that you saved over it.

5. In the winter, if the rice has grown to six inches (15 cm), you can allow ten ducks per quarter acre to graze in the field. If you notice any spots that are growing in thin, plant more seeds quickly. You don't want water accumulating during this time of year or it will freeze, so keep the rice drained.

6. In spring, check again for thin spots and sow more seeds if you need to.

7. In late spring, it is time to harvest the rye, barley, or other grain. You will have to walk on the rice to do this, but don't worry about it. Stack the grain to dry for about a week, and then thresh it.

8. Spread the threshed straw and husks on the field. If you have more than one plot, don't spread it in the same plot it was grown in. If you grew rye in one plot and oats in another, spread the oat straw in the rye field and vice versa.

9. By early summer, only rice is growing, and weeds. Now is the time to flood the paddy for about a week until the clover turns yellow (but isn't killed).

10. Throughout the summer, the field should always be at least half rice. Prepare the seeds of the grains for sowing, and pick a different grain to plant in the fall than you did the year before.

When to Harvest

Note: If fungus has got your grain, don't eat it and don't give it to animals. The mold can cause all kinds of problems.

Grain goes through various stages before it is ready to harvest. It starts out as

---

Bugs in the Grain

Dry ice: You will need one tablespoon of dry ice per five gallons (19 L) of grain. Get an airtight container, put the ice on the bottom, pour the grain on top, wait an hour, then seal the container. The grain produces carbon dioxide and kills the bugs.

Heat: You can't use this method for seed grain, but it works for edible grain. Spread a quarter inch (0.6 cm) on a pan and heat it in the oven 140°F (60°C) for 30 minutes.

---

a seed, sprouts, and becomes a seedling with a few short leaves. Then it becomes a *tiller* by sending out a couple of thicker shoots. The stem grows until it begins *booting*, or forming a head, at which point it is now *heading*. Heading continues until the head is completely formed, right up until it starts to flower. Flowering completes the pollination process, and that's when the grain enters the *milk* phase. This is when the kernel starts forming and you can squeeze out a milky fluid. Eventually, the kernel will dry out and will become mature during the *dough* stage. Ideally, grain should be harvested when it is in the *late dough* stage, when it is far past the milk stage but still able to be dented. When you let it dry, it will become dead ripe.

Grain yield in square feet:

Usually grain is measured in bushels, not pounds. One bushel is 64 pints or 35.2 liters. The average yield of an acre is 30 bushels (threshed and winnowed) or 1 ton. The following table gives an estimate of possible grain production for small acreage. But this is just an estimate. The

---

Type	Pounds per bushel	Sq. ft. to grow 1 bushel
Barley	48	900
Buckwheat	50	1,300
Field corn	56	500
Oats	32	600
Rye	56	1,500
Sorghum	50	600
Spelt	30	450
Wheat	60	1,000

pounds per bushel will vary year to year based on grain quality. Yield per square feet will also vary. These are general guidelines for measuring success and seeding rates. An acre is 43,560 square feet. A family of five eats about one thousand pounds of wheat in a year. That's sixteen bushels of wheat, which could be produced on 16,000 square feet, or about one-third of an acre.

About winter grain:

Winter grain is planted in September or later, so the seeds will "hibernate" during the cold weather and grow as soon as the weather warms up. It's better to have good snow cover for insulation—really cold weather will kill it. But don't let it stool (grow a stalk), because that can also kill it. Avoid it by planting later and letting animals graze over it a little. Plant winter wheat, barley, oats, or rye in early fall so it will sprout next spring.

Tillers and suckers:

A grain plant has a main stalk and head, but it also has secondary stalks, called tillers, that also produce grain. Corn also has these but they are called suckers because they don't produce anything. Sometimes planting less seed will cause more tillers, producing more grain than you would normally get, because there is less competition between plants.

To save grain seed:

You can save seed from your second crop of grain (unless you're using a hybrid). Your seed should be the best seed heads, unbroken and healthy. Let them dry in the shock for at least a month until they are totally dry. Then thresh them, and store the seed in a cool, dark place.

Storing Grain

You have to store grain in a rodent-proof container and keep hungry cats around (don't put it in a sack that has held seed grain because it can be highly poisonous). Keep the containers in a cool, dry place. It will keep for a year or more, until you grind it. Then use it immediately. Be smart about how you store it: it should be dry, in a bug-free area, without twigs or other debris, and mold-free.

Grain can get ergot fungus sometimes (but rarely). Don't ever eat moldy grain or feed it to your animals, because you can die and your animals can get sick. It turns the grain hard black and purple on the inside. It is caused when your grain gets damp. If your seed grain gets it, you should throw it out.

Ways to process grains:

Cracking: breaking the kernel in two or more pieces, done especially on corn

Crimping: flattening the kernel slightly, done especially with oats

Flaking: treating with heat and/or moisture, then flattening it

Grinding: forcing through rollers and screens

Rolling: smashing between rollers at different speeds with or without steaming

Saving: saving the wheat to grow as seed next year

Use a mortar and pestle, a hand grinder, or electric mill. For large jobs, you can either buy a big electric mill or a water-powered stone mill like the old 1800s type, which is beyond complicated. The best for a family is the electric mill because you only have to put it through once. With a hand grinder, you must crank the handle of the grinder for a long time, sift the grain, and put it through several times depending on the texture you want.

PLANT TABLES

Note: Just because a plant is listed as edible doesn't mean every species is safe to eat, or that the whole plant is edible. Some plant groups may only have one edible species and the rest are poison to humans, and some plant groups are only safe for animals. Read the plant notes for each one and make sure you have the right species.

The tables below are lists and uses of many kinds of edible and non-edible plants. They are suggested because of their variety of benefits, and when you take an inventory of your property you may find you already have some of them. Even though you have your hands full just growing food, using some of these plants can save you a lot of work in the long run. The following tables and guides are by no means complete. A variety of different climates and conditions are represented and should give you a starting list of species to experiment with. Contact your local agricultural extension to find out your best native plants, and use these lists as a guide to what to look for.

▲ Midway through the season, two beds have been cleared out and fertilized for the next crop.

Soil-Improving Plants

	Legumes	Nitrogen Fixers	Bare Soil Erosion	Stream Bank Erosion	Wetland Tolerant	Water Plants	Dryland Tolerant	Ground Cover	Edible
Acacia	X	X	X				X		X
Alder	X	X							
Alfalfa	X	X							X
Almond							X		X
Amaranth							X		X
American yellowwood	X	X							
Asparagus				X					X
Autumn olive	X	X	X						X
Bamboo		X	X	X					X
Beans	X	X							X
Bearberry								X	X
Black locust	X	X	X				X		
Black tupelo			X	X	X				
Bunchberry				X	X			X	X
Carob	X	X					X		X
Cattail						X			X

The Ultimate Guide to Urban Farming

Plants	Legumes	Nitrogen Fixers	Bare Soil Erosion	Stream Bank Erosion	Wetland Tolerant	Water Plants	Dryland Tolerant	Ground Cover	Edible
Ceanothus	X	X							X
Clover	X	X						X	X
Cranberry						X			X
Daylily			X		X				X
Duck potato						X			X
Duckweed						X			X
Fenugreek	X	X							X
Fig							X		X
Flowering dogwood					X			X	X
Gliricidia	X	X							X
Groundnut	X	X							X
Hawthorn			X				X		X
Hairy vetch	X	X	X						
Holm oak							X		
Honey locust	X	X					X		X
Jasmine					X				X
Japanese pagoda tree	X	X							X

	Legumes	Nitrogen Fixers	Bare Soil Erosion	Stream Bank Erosion	Wetland Tolerant	Water Plants	Dryland Tolerant	Ground Cover	Edible
Jujube							X		X
Lavender							X		
Leucaena	X	X		X				X	X
Lemongrass				X					X
Licorice	X	X							X
Lotus						X			X
Lowbush blueberry								X	X
Lupine	X	X						X	X
Mesquite	X	X					X		
Mint					X			X	X
Mulberry							X		X
Nasturtium								X	X
Nettle								X	X
N.Z. spinach							X	X	X
Olive							X		X
Palo Verde	X	X	X				X		X
Paulownia			X						X
Pea	X	X							X

	Legumes	Nitrogen Fixers	Bare Soil Erosion	Stream Bank Erosion	Wetland Tolerant	Water Plants	Dryland Tolerant	Ground Cover	Edible
Peanut	X	X							X
Persian ground ivy								X	
Pigeon pea	X	X					X		X
Pomegranate							X		X
Poplar				X	X				
Prickly pear							X		X
Quinoa							X		X
Raspberry								X	X
Rice						X			X
Rosemary							X		X
Rush						X			
Russian olive	X	X	X				X		X
Siberian pea shrub	X	X					X		X
Sesbania		X			X		X		X
Sorghum							X		X
Stone pine							X		X
Strawberry								X	X

	Legumes	Nitrogen Fixers	Bare Soil Erosion	Stream Bank Erosion	Wetland Tolerant	Water Plants	Dryland Tolerant	Ground Cover	Edible
Sunflower									x
Sweetfern	x	x					x	x	x
Sweet potato								x	x
Sweet woodruff					x				x
Taro					x				x
Tagasaste	x	x					x		x
Thyme								x	x
Vetch	x	x							
Violet								x	x
Watercress				x	x	x			x
Wild ginger					x			x	x
Wild rice						x			x
Willow				x	x	x			x
Yarrow								x	

Plants by Harvest Time

Less than 2 months to harvest	2–3 months to harvest	3–6 months to harvest	6+ months to harvest
Arugula	Green beans	Artichoke	Garlic
Greens	Beets	Dry beans	
Basil	Broccoli	Brussels sprouts	
Broccoli rabe	Carrots	Cabbage	
Chard	Cauliflower	Corn	
Cilantro	Celery	Leeks	
Dill	Eggplant	Melons	
Green onions	Kale	Onions	
Radish	Peas	Winter squash	
	Peppers	Sweet potatoes	
	Potatoes		
	Shallots		
	Summer squash		
	Tomatoes		
	Turnips		

Plants by Climate Tolerance

Temperate	Cold Hardy	Warm
Artichoke	Anise	Basil
Arugula	Arugula	Beans
Asian greens	Asian greens	Chayote
Basil	Basil	Corn
Beans	Beets	Cucumber
Beets	Broccoli	Dill
Bitter greens	Brussels sprouts	Eggplant
Broccoli	Cabbage	Melons
Brussels sprouts	Caraway	Okra
Cabbage	Carrots	Peppers
Cardoon	Celery	Potatoes
Carrots	Chamomile	Squash
Celery	Cherry tomato	Tomatoes
Chard	Chervil	
Cilantro	Chive	
Collards	Cilantro	

Temperate	Cold Hardy	Warm
Fennel	Collards	
Garlic	Cucumber	
Green onions	Garlic	
Kale	Lettuce	
Leeks	Mint	
Lettuce	Nasturtium	
Mustard greens	Onions	
Onions	Parsley	
Parsley	Radish	
Shallots	Rosemary	
Radish	Sage	
Rutabaga	Swiss chard	
Spinach	Tarragon	
Turnips	Thyme	
	Turnip tops	

ALPHABETICAL PLANT AND HARVEST GUIDE

Note: Harvest descriptions include tool requirements. Washing boxes are trays with holes for dunking edible greens. Bins are solid boxes for carrying veggies.

Alfalfa

Usage: Bee forage, edible, leguminous nitrogen fixer

Species: Alfalfa (*Medicago sativa*)

Growing: Alfalfa is the forage and hay of choice for livestock, although too much of a good thing can be fatal if an animal is suddenly switched to it exclusively. People eat alfalfa sprouts, too, which are incredibly healthy. It is a perennial legume that grows well in well-drained loam. Wait until the alfalfa just starts to bloom before you allow the animals in to forage or you cut it for hay, and don't let it get shorter than two inches (5 cm). Let it recover for another six weeks, then do it again. Make sure you cut it a month before the first frost.

Almond

Usage: Nut, bee forage, dryland tolerant

Species: Sweet Almond (*Prunus dulcis*)

Growing: There are many, many varieties of almond, and most of the North American ones are grown in California, where they enjoy the warm, sunny climate. The sweets are the edible kind, and it's just not worth it to grow the bitter variety, which you can't eat. Some varieties are not self-pollinated, so you need a few trees. It generally takes five years to see any kind of real production. They only grow around fifteen to thirty feet (5–9 m) tall.

▲ This large greenhouse built of PVC and wood is not very durable, but it's very inexpensive and can grow tomatoes and other warm-temperature crops.

They like plenty of water since they are usually grown in dry and hot areas, and may be susceptible to peach leaf curl, a fungus, which turns the leaves brown and curls them. Remove the infected leaves and burn them. The tree will produce small green hulls that will begin to dry and drop off the tree, although you will probably have to pick some of them. Remove the almond from the husk and let them dry. When they are fully dry, which takes a few days, the nut will rattle inside the shell. They can be eaten raw or roasted.

Amaranth

Usage: Edible leaves and seeds, dryland tolerant
Species: Purple Amaranth (*Amaranthus blitum*), Red Amaranth (*Amaranthus cruentus*)
Growing: Amaranth is treated like a grain, although it's not a grass. It has broad leaves, which
   are some of the healthiest leaves you can eat and full of protein. Amaranth is dryland
   tolerant and enjoys well-drained soil and full sun. It grows best in the southern US. For a
   similar plant that can be grown in northern areas, choose quinoa. To harvest the seeds
   for grain, you must rub the flower heads to see if the seeds fall out easily. Sometimes
   the plants continue flowering long after the seeds are formed, so flowers aren't a very
   good indication of readiness. Harvest the seeds by shaking and rubbing the heads into a
   bucket. You will need to thresh away the hulls like any other grain. The simplest way to do
   this is by rubbing the flower heads on a large screen and then using a fan to blow the chaff
   away. Use the same screen trays to dry the seeds, either in the sun or near the wood stove.
   Store in a cool, dry place. Amaranth can be cooked just like rice, or ground into flour and
   added to bread.

Apple

Usage: Bee forage, edible fruit, wood and timber
Species: Apple (*Malus domestica*)

Plants

Growing: There are more than seven thousand kinds of apples, and they grow in every climate. Some are for eating, some are for cooking, and some are for drinking. They must be grown in groups because they are not self-pollinating and are often grafted for this reason. They are susceptible to pests and diseases, and if not pruned, will grow to amazing and unpickable sizes. To counteract this, the home grower can choose a dwarf variety, which is much easier to manage.

Apricot

Usage: Bee forage, edible fruit

Species: Apricot (*Prunus armeniaca*)

Growing: Apricots are hardier to cold temperatures than peaches, which means they can grow in cold climates, but they are also susceptible to spring frosts, which can kill the early flowers. They are sensitive to soil type and need fertile, well-drained soil. You will need several trees to grow them, as they are cross-pollinated. They won't bear fruit for two or three years, but you can increase production by pruning after the second year to remove any crossed or rubbing branches. Keep the pruning simple and to a minimum, and do it on a hot day in the summer, to prevent sickness. They are not drought-tolerant and may sometimes need some watering, but their roots should never sit in water for too long.

Arugula

Usage: Edible leaves

Species: *Erica sativa* (Brassicaceae or cabbage family)

Growing: Arugula can be planted in an early garden without cover crops and not much plant debris because of the tiny seedlings. Late arugula can follow a spring-planted lettuce crop. It does not need much weeding, as it crowds out weeds, but it is susceptible to flea beetles. Use a floating row cover to avoid flea beetle damage in your earliest plantings. Fall plantings don't usually need that kind of protection. It is a short season crop that can be followed by lettuce or spinach.

Arugula harvest grid:

Yield	Harvest Time	Tools	Harvest Readiness	Washing	Processing	Storage
¼ lb. per row foot	5 bins per hour @ 12 lb.	Knives, washing boxes, salad spinner	3–4 inches tall for salad, 5–6 for cooking, deep green, leave 1 inch for regrowth	12 bins per hour; dunk twice in ice water, no longer than 15 seconds	Pull out weeds and yellow leaves	Remove excess water and keep cold

Asparagus

Usage: Stream bank erosion control, edible roots/shoots

Species: Asparagus (*Asparagus officinalis*)

Growing: A perennial that enjoys full sun, asparagus is delicious but finicky and not able to compete with other plants. It needs to grow alone, with the exception of tomatoes, which repel asparagus beetles. Some people are allergic to the raw shoots, but they are better cooked anyway. Asparagus is a high-nutrition food and worth going to a little trouble for. It is cold hardy and prefers well-drained soil. Traditionally, asparagus is grown in a trench about a foot deep (30 cm), with a bed of three inches (8 cm) of composted manure or mushrooms at the bottom. Adding manure tea after it is planted is also beneficial. It needs regular watering and weeding, but with proper care will keep growing for fifteen years. You can begin harvesting asparagus spears the year after planting. Midspring is the best planting season for this vegetable because the weather is a bit warmer and drier. Pick only spears that are between seven and nine inches (18–23 cm) tall with closed tips. You can only pick them for four weeks, and you must make sure to pick them before the tips fan out, as this makes them susceptible to those pesky beetles. When picked, throw them in ice water to chill them, and then put them in a plastic bag in the fridge, where they will last a couple of weeks. They have deep and quick-growing roots that can be used to stabilize stream bank soil.

Autumn Olive

Usage: Bare soil erosion control, edible fruit, bee attraction, windbreak, leguminous nitrogen fixer

Species: Autumn Olive (*Elaegnus umbellata*)

Growing: This is a hardy, small tree that produces a small, nutritious fruit that can be eaten fresh, dried, or turned into jam. The tree also improves the soil by fixing nitrogen. Since it grows in most soil conditions, it can stabilize eroding bare soil and act as a windbreak. Despite its value and versatility, in North America it is often considered an invasive plant or weed because it is difficult to remove. Once established, it will continue to grow back from the roots and cause trouble for other species. Because of this, it should only be grown in places where it already grows, which is mainly in eastern North America.

Avocado

Usage: Edible fruit

Species: Avocado (*Persea americana*)

Growing: The avocado grows seventy feet tall and needs a frost-free climate. They are not tolerant to wind and need deep, well-drained soil. They are also susceptible to diseases. The fruit must ripen after it has been picked. Typically, they simply begin to fall off the tree and ripen on the ground. Pick and store them at room temperature until they are ripe, which usually only takes a few days. They are eaten raw in a variety of dishes, including desserts. They have high amounts of healthy fat and are very nutritious. Beware: the plant and pit are toxic to animals.

Azolla

(see Duckweed)

Bamboo

Usage: Bare soil erosion control, stream bank erosion control, edible roots and shoots, wood and timber

Edible Species: Sweetshoot Bamboo (*Phyllostachys dulcis*), Stone Bamboo (*Phyllostachys nuda*), Blue-green Claucous Bamboo (*Phyllostachys viridiglaucescens*), Temple Bamboo (*Semiarundinaria fastuosa*)

Growing: Bamboo enjoys partial shade to full sun, and can be used for almost anything. They can quickly take over a garden and so they need to be contained. With that said, they are very hardy and don't need any work. The shoots pop out of the ground on their own, but they need to be cooked. Choose shoots that are heavy and firm with a thick outer skin. Peel that skin off and cut off the root end. Put it in a pot of water and bring it to a boil, then turn it down to a simmer for about an hour or until it is soft through the middle. Turn off the heat and let it just sit there until it is cool. There is probably some extra skin still on it that needs to be peeled off, but it should be ready to eat. Keep the cooking water and use it to store the bamboo in, so it won't dry out. The shoots can be eaten cold on rice, or used in stir-fry.

Basil

Usage: Edible herb

Species: *Ocimum basilicum* (Lamiaceae or mint family)

Growing: This tropical herb comes in three main varieties: Genovese, Large Leaf & Sweet Thai. It should not be planted after basil or cut flowers. It does well on clean ground and it can be seeded or transplanted. Transplants should not be older than four weeks. Basil needs to be harvested before the first frost. Black spots indicate downey mildew, which can only be prevented by growing the basil in a greenhouse or high tunnel. Early garden basil can be planted with mulch and hoop covers.

Basil harvest grid:

Yield	Harvest Time	Tools	Harvest Readiness	Washing	Processing	Storage
⅓ lb. per foot	6–8 bins per hour	Scissors, knives, bins	Plants 10–15 inches tall, deep green, no holes, black spots, or flowering stems	Do not wash	Pull out weeds and yellow leaves, flowers, black spots	Keep cool but not in cooler with good air circulation; do not get wet

The Ultimate Guide to Urban Farming

Beans

Usage: Edible, leguminous nitrogen fixer, climbing vine
Species: Common Bean (*Phaseolus vulgaris*), Scarlet Runner Bean (*Phaseolus coccineus*),
   Winged Bean (*Psophocarpus tetragonolobus*)
Growing: There are three kinds of beans: snap, shell, and dry. The snap varieties can be
   harvested when the pod is developed but the beans are small, and they are eaten
   whole. The shell varieties must be picked when the beans are more developed but not
   dried. Dry varieties are picked when the beans rattle in the shell. They can grow in most
   soil types and need something to climb on. Dry beans need about three to four months
   to reach an acceptable level of dryness, which usually happens by early September.
   Even then, let them sit on a drying rack in a warm, dry place for a couple of days before
   shelling. Remove the shells, blow the shells (chaff) away and remove any other debris or
   bad beans. Bad beans are moldy or discolored or have holes in them. They should be
   so dry that you can't dent them with a fingernail. Other kinds of beans can be harvested
   continuously throughout the season, and in fact, the more you pick, the more will grow.
   The Scarlet Runner Bean is unique in that it grows a big tuberous root and tolerates
   colder temperatures, but the growing is pretty much the same. Keep picking them
   as they grow. The Winged Bean is a quick-growing vine that needs to be given lots of
   space to climb and a strong trellis. It prefers hot, humid places but there are varieties
   that can grow just about anywhere, and the entire plant is edible. The pods are picked
   continuously, the young leaves can be used like spinach, and the root can be cooked
   like a potato. The root is full of protein. The beans can also be dried.

Beans harvest (snap) grid:

Yield	Harvest Time	Tools	Harvest Readiness	Washing	Processing	Storage
⅓ lb. per foot	Up to 50 bins per hour	Boxes	Before pod is filled with beans	Do not wash	Remove debris	Keep cool but not in cooler; good air circulation; do not get wet

Bearberry

Usage: Edible fruit, ground cover
Species: Red Bearberry (*Arctostaphylos rubra*), Alpine Bearberry (*Arctostaphlos alpine*)
Growing: Bearberry is a small evergreen shrub that is used as a ground cover. It has
   the additional benefit of producing edible berries. It enjoys full sun, tolerates most
   soil types, and is drought-resistant. It does prefer a cooler climate and is beautiful
   for adding greenery in the winter. For the first few years, it will grow very slowly.
   The berries aren't particularly nice to eat. Traditionally, they were used for herbal
   medicine—dried and ground for mixing into other foods, or turned into jam.

Bee Balm

Usage: Bee forage, edible

Species: Bee Balm (*Monarda fistulosa*), Scarlet Bee Balm (*Monarda didyma*)

Growing: Bee Balm has many traditional medicinal uses and is easy to grow. It thrives in most areas and simply needs regular watering. Cut them back once a year to keep them between four and eight inches (10–20 cm) tall. They can become aggressive spreaders, so dividing the plant every two to three years will keep them in check and also help keep them healthy.

Beet

Usage: Edible roots and leaves

Species: Beetroot (*Beta vulgaris*), Swiss Chard (*Beta cicla*)

Growing: Beets are incredibly nutritious and can be eaten raw or cooked. While the colorful root bulb is the more common food, some people use the beet greens in salad from time to time. Don't eat too much of those, as it can make you lose calcium in your body. They are cold hardy and need fertile soil. Harvest when they reach at least one and a half inches (3.8 cm) in diameter. They can be stored in live storage or pickled. Swiss chard is in the same family but doesn't form a fleshy root. Instead, the leaves are harvested as they grow, like lettuce. These highly nutritious greens can be eaten raw when young, but it is more common to cook them like spinach. Both of these varieties can be grown in the greenhouse over the winter.

Beet harvest (greens) grid:

Yield	Harvest Time	Tools	Harvest Readiness	Washing	Processing	Storage
0.2 lb. per foot	5–6 bins per hour	Scissors, knives, boxes, salad spinner	Greens 3–4 inches tall for salad, 5–7 inches for cooking	Dunk boxes twice in ice water; do not leave in water for more than 15 seconds	Pull out weeds, yellow leaves, and black spots	Store in cooler, remove excess water

Beet harvest (roots) grid:

Yield	Harvest Time	Tools	Harvest Readiness	Washing	Processing	Storage
⅓ bunch per foot	30–50 lb. bunches per hour	Bins	2–3 inches wide, round, unblemished leave beets with black marks	Do not wash	Pull out weeds and yellow leaves; dust off excess dirt	Store in cooler

Blackberry

Usage: Edible fruit, bee forage

Species: Blackberry (*Rubus spp.*)

Growing: Blackberries grow wild everywhere in North America and the sweet berries make excellent jam, desserts, and snacks. They are very hardy and easy to grow. With a trellis they can also act as a useful hedge. Simply pick the berries when they are ripe.

Blueberry

Usage: Edible fruit, bee forage, ground cover

Species: Highbush Blueberry (*Vaccinium corymbosum*), Rabbiteye Blueberry (*Vaccinium asbei*), Lowbush Blueberry (*Vaccinium angustifolium*), Creeping Blueberry (*Vaccinium crassifolium*)

Growing: Blueberries enjoy full sun and grow to be six to twelve feet high (2–4 m) and six to twelve feet wide. They need highly acidic, well-drained soil. The Highbush varieties are hardier and also more common, but the Rabbiteyes produce more fruit. Lowbushes don't make the greatest groundcover like the other varieties, but have the best fruit. The fruit can be eaten raw, sun-dried, or as dessert and jam, and is very nutritious. They do better with regular watering. The picking season begins sometime around May or June and lasts until the end of summer.

Borage

Usage: Bee forage, edible

Species: Common Borage (*Borago officinalis*)

Growing: Borage is a valuable medicinal plant and is also a nutritious addition to salads. It also grows well as a companion plant. They are a hardy herb and bloom most of the season and reseed themselves. The leaves can be picked as needed.

Broccoli

Uses: Edible

Species: Broccoli rabe (*Brassica rapa*), Broccoli (*Brassica oleracea*)

Growing : Broccoli, as a brassica, should not follow another brassica crop. It needs a good amount of compost, and it doesn't like too much heat. Purple leaves indicate a phosphorus deficiency, which can be remedied with a spray. Transplants shouldn't be more than four weeks old, but they should have four true leaves. Broccoli is susceptible to flea beetles and cabbage worms. A floating row cover can help prevent those kinds of pests.

Broccoli harvest grid:

Yield	Harvest Time	Tools	Harvest Readiness	Washing	Processing	Storage
⅓ bunch per row foot	200 heads per hour	Scissors, bins	Tightly bunched, green florets, no black spots	Soak in ice water for 30 minutes	Remove debris and leaves	Store in cooler with ice

Broccoli rabe harvest grid:

Yield	Harvest Time	Tools	Harvest Readiness	Washing	Processing	Storage
⅓ lb. per row	50–70 bunches per hour	Knives, box	Greens 12–18 inches tall with small heads, 8–10 inches for young leaves, stem should have no white in core	Dunk in ice water for 30 seconds, dunk in second bin for 30 seconds (while you do next box)	Remove dead leaves	Store in cooler

Brussels Sprouts

Usage: Edible leaves and heads

Species: Brussels sprouts (*Brassica oleracea*)

Growing: As a brassica, Brussels sprouts should not follow another cabbage family crop. They do well in raised beds with lots of compost and can be transplanted to give them a head start. The transplants should not be more than five weeks old. Keep them well weeded. Pests include flea beatles, cabbage worms, and cabbage loopers, which can be prevented with row covers and organic sprays.

Brussels Sprouts harvest grid:

Yield	Harvest Time	Tools	Harvest Readiness	Washing	Processing	Storage
¼ stalk per row foot	60 stalks per hour	Knives, box	Sprouts are firm, green, and round, outer leaves can have some black spots	Do not wash	Remove dead leaves and diseased sprouts	Store in cool area with high humidity and good air circulation

Buckwheat

Usage: Bee forage, edible, cover crop

Species: Common Buckwheat (*Fagopyrum esculentum*)

Growing: Although buckwheat is eaten as a grain, it is not a grass or a cereal. It grows very quickly, ripening in about ten to thirteen weeks, and so it makes a great crop for a cooler climate. It does not tolerate frost and should be planted late in the season to ripen in early September. Buckwheat requires regular watering and soil that does not have to be great, but at least loose. Due to its short growing season, it should be double-cropped with another grain, such as winter wheat, oats, or flax. If your earlier grain crops failed, it may also be possible to plant buckwheat and quickly raise an emergency food source. As a cover crop, it effectively chokes out weeds. The seeds have a hard outer hull, which must be removed, and the seed can be used whole or ground as flour. Traditionally, buckwheat has been used for noodles and pancakes. It has no gluten and so can only be used in small amounts when making bread. The hulls are used as a filling for pillows or other items. To harvest, cut the buckwheat down when it is almost all brown, but still has a couple of green leaves or flowers. Cut it as you would a grain, by mowing and swathing. Thresh the grain, remove the chaff, and then allow the seeds to dry. The seeds still have hard hulls on them, but they need to be very dry to be removed (and you don't want them to get moldy). You can do this by laying them out on your drying trays in the hot sun or using your dehydrator. Then you can use a grain mill to dehull the seeds. You must use the largest setting and run them through a few times to crack open all the hulls. Sift them until you have removed all the hulls, which you can save for pillows or other projects.

Bunchberry

Usage: Stream bank erosion control, wetland tolerant, edible fruit, ground cover
Species: Canadian Bunchberry (*Cornus canadensis*), Bunchberry (*Cornus suecica*)
Growing: Bunchberries are perennials that like shady, cool, and moist places. It is found
   in northern areas and grows around stumps and mossy areas that stay wet most of
   the time. It needs to be in a spot where the soil never gets warmer than 65°F (18°C)
   and hardly ever sees the sun. It enjoys soil full of rotting wood, which can be added
   as mulch. The berries are edible but are a little bland. They taste best used in
   sauce, jams, and pudding. The berries also help thicken up jam so you don't need
   to add pectin. The berry tea is a traditional herbal remedy.

Cabbage

Usage: Edible
Species: Chinese cabbage (*Brassica rapa*), Green and Red (*Brassica oleracea*)
Growing: Cabbage, like other brassicas, should not follow another brassica crop. They need
   a lot of compost and nitrogen-rich fertilizer. They do well in raised beds. They can be
   transplanted no more than five weeks old, before they start to get "leggy," meaning long
   and skinny with few leaves. Cabbage is susceptible to flea beetles, cabbage worms,
   and cabbage loopers, which can be prevented with row covers and organic sprays.

Carob

Usage: Seedpod and sugar source, edible, dryland tolerant, leguminous nitrogen fixer
Species: Carob Tree (*Ceratonia siliqua*)
Growing: Carob is a large evergreen shrub that grows to fifty feet (15.2 m) and prefers
   hot climates. In the Mediterranean and Australia, it is grown commercially, but
   in North America it could be grown in the southern states. It is not frost hardy,
   although a full-grown tree could withstand 20°F (7°C) if the cold temperature does
   not occur during blooming, which would prevent any fruit from growing. It won't

tolerate any kind of heavy rain until the end of the growing season. It can grow in any soil and is tolerant to drought. The carob pods are sweet and have many uses. To harvest them, shake the tree with a long pole before the winter rainy season starts. Catch the pods on a sheet or tarp on the ground. Allow the pods to dry in the sun for a couple of days until the seeds rattle in the pod. The pods can be crushed to break up the seeds and fed to animals. But animals should not eat too much of it, as it will stunt their growth. Chickens cannot eat them at all. When consumed by humans, they are most often ground into varying consistencies for use in making syrup, jam, fine flour, and a coffee substitute.

Cabbage harvest grid:

Yield	Harvest Time	Tools	Harvest Readiness	Washing	Processing	Storage
¾ head per row foot	200 heads per hour	Knives, bins	Firm heads, no damage from insects or black rot	Do not wash	Remove dead and damaged leaves	Store in cool area with high humidity and good air circulation

Carrot

Usage: Edible roots and shoots

Species: Carrot (*Daucus carota*)

Growing: Carrots are a familiar root vegetable that come in orange, white, purple, red and yellow. The greens are also edible. In planting, they are a useful companion plant, especially for tomatoes. Carrots enjoy partial shade—rather than full sun—in loose soil. They take at least four months to grow.

Carrot harvest grid:

Yield	Harvest Time	Tools	Harvest Readiness	Washing	Processing	Storage
Half bunch per row foot	80–100 heads per hour	Boxes	Healthy green tops, sweet, 5–6 inches long	Soak while rinsing 3 bunches (5–10 in each bunch) at a time, soak in cold water 5 minutes then store	Remove dead and damaged leaves	Store in cooler

Cauliflower

Usage: Edible heads

Species: Cauliflower (*Brassica oleracea*)

Growing: Like all other brassicas, cauliflower should not follow another brassica. It does
   well in raised beds with lots of compost, but may need additional nitrogen-rich
   fertilizer; plant it after beans to help with this. Transplant no later than five weeks.
   Protect against flea beetles, cabbage worms, and cabbage loopers. One week before
   harvest, fold the large leaves over the heads and tie them with a rubber band. Some
   varieties are self-blanching and don't need this. After harvest you can turn the crop
   residue into the soil.

Cauliflower harvest grid:

Yield	Harvest Time	Tools	Harvest Readiness	Washing	Processing	Storage
¼ marketable head per row foot	40 heads per hour	Knives, boxes	Firm, white heads, no worms	Do not wash	Remove dead and damaged leaves and keep some of the leaves around the head	Store in a cool place with high humidity

Cattail

Usage: Water plants, edible

Species: Narrowleaf Cattail (*Typha angustifolia*), Southern Cattail (*Typha domingensis*),
   Broadleaf Cattail (*Typha latifolia*)

Growing: All parts of the cattail are edible. The young shoots are cut in the spring when
   they are at least four inches (10 cm) long. The roots can be boiled like potatoes, the
   flower stalks can be boiled or steamed like corn, and the pollen works as a flour
   substitute. They grow everywhere in the world, and the cattails can also be woven into
   useful items like mats and baskets. They are perennial water plants and, unless kept in
   check, can take over a pond. They are hardy and easy to grow from cuttings or seeds.

Ceanothus

Usage: Leguminous nitrogen fixer, ground cover, edible

Species: New Jersey Tea (*Ceanothus americanus*), Snowbrush Ceanothus (*Ceanothus
   velutinus*), Maritime Ceanothus (*Ceanothus maritimus*)

Growing: Ceanothus is an evergreen species. However, in very cold climates a few species
   are deciduous. The leaves are edible and have traditionally been used in herbal
   medicine and teas. There are many varieties of Ceanothus, all very hardy, and so it is
   probably a good idea to just pick your native species.

Cedar

Usage: Pest control, wood and timber

Species: Deodar Cedar (*Cedrus deodara*), Lebanon Cedar (*Cedrus libani*), Cyprus Cedar (*Cedrus brevifolia*), Atlast Cedar (*Cedrus atlantica*)

Growing: Cedars grow in the mountains where the winters aren't too severe but still get snow or monsoon rains. They grow very big and should not be planted near any overhanging obstruction or over buried pipes or cables. They enjoy partial shade to full sun. The wood is resistant to rot and pests and has a pleasant scent.

Celery

Usage: Edible stalk

Species: Celery (*Apium graveleolens*)

Growing: Celery is a fresh, tasty vegetable that is eaten for its stalk and leaves. It should not follow other umbellifers (the carrot family). It needs lots of compost and can be grown in raised beds. A common problem is lack of nitrogen, calcium, and boron so the more organic, the better.

Celery harvest grid:

Yield	Harvest Time	Tools	Harvest Readiness	Washing	Processing	Storage
⅔ piece per row foot	175 pounds per hour	Knives and bins	Roots should be 3–6 inches around, usually late fall	Do not wash until ready to deliver, then spray off dirt	Remove dead and damaged leaves	Store in cool place with high humidity

Chard

Usage: Edible leaves and stalks

Species: Swiss Chard (*Beta vulgaris var. cicla*)

Growing: Swiss chard is a hardy spinach or kale-type vegetable that can grow to amazing sizes. It should not follow after spinach or beets, and it needs lots of compost. Don't let them get leggy before transplanting seedlings—probably no older than four weeks. After harvest the roots will regrow and act as a weed—either pull them up or plant a cover crop after.

Cherry

Usage: Edible fruit, wood and timber, bee forage

Bush Species: Mongolian Bush Cherry (*Prunus fruticosa*), Japanese Bush Cherry (*Prunus japonica*), Nanking Cherry (*Prunus tomentosa*), Choke Cherry (*Prunus virginiana*)

Growing: There are hundreds of cherry varieties, but the bush varieties may be the most useful as they don't take up much space, growing three to eight feet (1–2.5 m) high and up to eight feet wide. There are even dwarf bush varieties that take up even less space, and are even more resistant to pests and diseases. Mongolians are hardy, but sourer. Nankings are sweeter and grow anywhere, including dry places. Cherries enjoy long, warm summers, but need to get cold in the winter, which makes them perfect for a temperate climate. They need fertile, well-drained soil. Sour cherries are often self-pollinating but sweet cherries need to be planted in groups. Pick the cherries as they ripen, and eat raw, dry them, or make them into pies.

Card harvest grid:

Yield	Harvest Time	Tools	Harvest Readiness	Washing	Processing	Storage
1 bunch per row foot	80–120 bunches per hour	Knives and boxes	Plant is 15–18 inches tall, bottom leaves large enough to bunch	Dunk in water to clean, then dunk in ice water for 1 minute to cool	Remove dead and damaged leaves	Store in cooler

Chestnut

Usage: Edible nut, wood and timber
Species: Chinese Chestnut (*Castanea mollissima*)
Growing: The chestnut grows to be eighty feet tall and is self-pollinating. It enjoys well-drained soil of any type, is tolerant to cold, and prefers partial shade to full sun. The chestnut can be eaten raw, but it is much better cooked and can be used as a staple like potatoes. When the nuts are ready to harvest, the spiny burs will begin falling to the ground, which usually happens in autumn. Most of the burs will split open. Gather those up and remove the nuts, throwing out any that are damaged or have holes in them. Put them in an airtight container or freeze them.

Chicory

Usage: Edible roots and shoots, bee forage
Species: Chicory (*Cichorium intybus*)
Growing: Chicory tastes a little bitter, but is very nutritious. It goes well with other baby greens in salads. It enjoys partial shade to full sun, and some varieties are perennial. They are often pulled out of yards as a weed because they can take care of themselves very well. Radicchio and Belgian endive are in the same family. They make excellent animal forage as the chicory kills worms and is easy to digest. It also makes a useful herbal remedy and coffee substitute.

Cilantro

Usage: Edible roots and shoots
Species: Cilantro (*Eryngium foedum*)

Growing: Cilantro can't be planted after other umbillifers (carrot family), potatoes, or curcubits. It does do well after lettuce and greens. If you use lots of compost, it only takes thirty–forty days to harvest.

Cilantro harvest grid:

Yield	Harvest Time	Tools	Harvest Readiness	Washing	Processing	Storage
1 bunch per row foot	3–4 boxes per hour	Knives and boxes	4–5 inches tall, green leaves with no yellow	Dunk in first tub for 30 seconds, drain, dunk in ice water while washing next box	Remove dead and damaged leaves	Store in cool, humid place

Clover

Usage: Bee forage, ground cover, edible

Species: White Clover (*Trifolium repens*), Bush Clover (*Lespedeza bicolor*), Red Clover (*Trifolium pretense*)

Growing: Clover enjoys partial shade to full sun and is a helpful nitrogen-fixing ground cover. It is also a valuable animal fodder and very hardy to being stomped on. Clover is easy to grow and spreads to form a mat.

Comfrey

Usage: Bee forage

Species: Large-flowered Comfrey (*Symphytum grandiflorum*), Russian Comfrey (*Symphytum x uplandicum*)

Growing: Comfrey is easy to grow in most climates, with the option to cut them back as mulch a few times per year. Once you have decided to grow them, you will never be able to get rid of them, so be sure you want them. It is mentioned several times in this book as a fertilizer because comfrey adds many nutrients to the soil. It prefers lots of nitrogen and so a bed of manure is a great home for it. Comfrey is also useful as a topical medicinal herb, but can cause liver damage when ingested in quantities.

Cranberry

Usage: Water plants, edible fruit, bee forage

Species: Common Cranberry (*Vaccinium oxycoccos*), Small Cranberry (*Vaccinum microcarpum*), Large Cranberry (*Vaccinium macrocarpon*), Southern Mountain Cranberry (*Oycoccus erythocarupus*)

Growing: Cranberries are evergreen dwarf shrubs and vines that grow a sour, red berry. The cranberries can be made into sauce, juice, jams, and dried fruit by adding sweeteners to make them delicious. They are grown similar to rice, in a paddy surrounded by a barrier. While the plants require regular irrigation, the field is not flooded until the end of the year. While big producers float the berries to harvest them, small producers will want to dry pick the berries to avoid damaging the crop and use less equipment. Then the beds are flooded so that they don't freeze over in the winter. Every four years, a thin layer of sand can be spread on the top to help control pests. The cranberries are stored in well-ventilated crates in a cool, dark place until they are processed.

Cucumber

Usage: Edible fruit
Species: Cucumber (*Cucumis sativus*)
Growing: Cucumbers are a cool, refreshing fruit that can be turned into pickles. They should not be grown after other curcurbits or tomatoes. Cucumbers need a lot of nitrogen-rich fertilizer. Row covers protect against striped cucumber beetles, but should be removed during flowering for varieties that need pollinators.

Cucumber harvest grid:

Yield	Harvest Time	Tools	Harvest Readiness	Washing	Processing	Storage
7 cucumbers per row foot	6 bushels per hour	Boxes	4–6 inches long, firm, dark green	Do not wash	Skin is free of scab	Store in cool place with high humidity

Currant

Usage: Edible fruit
Species: Black Currant (*Ribes nigrum*), Clove Currant (*Ribes odoratum*), Red and White Currant (*Ribes Silvestre*)
Growing: In Europe, these are as common as blueberries, but in many places in North America they are illegal because some species harbor a pest that kills white pines. They are very hardy to cold temperatures and can grow in most climates. Their one weakness is a susceptibility to rust, but many varieties are resistant, especially Red and White Currants. They are generally easy to grow, but check on your state's restrictions.

Dandelion

Usage: Edible plant, bee forage
Species: Dandelion (*Taraxacum officinale*)
Growing: Dandelions grow spontaneously on their own in everyone's yard and don't need any care at all. They are also extremely nutritious and the entire plant can be eaten. The flowers are made into wine or added to pancakes, the roots are brewed into a tea that tastes like coffee, and the leaves are added to salads. They can be grown in raised beds with lots of compost and can be harvested twice if they are well-weeded.

Daisy

Usage: Pest control

Species: Pyretheum Daisy (*Chrysanthemum cinerariifolium*)

Growing: These flowers prefer dry and somewhat sandy soils. They need to be weeded
   but since they are so hardy against pests, they should not have many other problems.
   To harvest them for use as a natural pesticide, wait until a warm, sunny day when the
   flowers have been already open for a few days. Dry them by hanging upside down or
   removing the heads and drying them in the sun. The flowers must be stored whole in a
   dark, airtight container. When you are ready to use them, grind them into a fine powder
   and either dust or spray them (mixed with some water) on the plants that are affected.
   The insects should die fairly quickly, without harm to humans.

Daylily

Usage: Bare soil erosion control, edible roots and shoots, bee forage

Species: Tawney Daylily (*Hemerocallis fulva*)

Growing: Daylilies enjoy partial shade to full sun. The entire plant can be eaten, including
   the flowers and roots. They are a little too large and expansive to be grown near other
   plants, but since they require little care, they can be planted in the outer zones. The
   Tawney Daylily is wetland tolerant. The edible parts of the daylily are the flower buds
   and flowers, which are most often used in Asian stir-fry dishes. Cut off the base with the
   ovary and use like mushrooms. Only the cultivated varieties of daylily are edible; others
   are toxic, so make sure you have the right one.

Dill

Usage: Pest control, edible

Species: Dill (*Anethum graveolens*)

Growing: A perennial herb that enjoys full sun. It grows in just about any climate but prefers
   fertile, well-drained soil. The leaves can be used fresh, and the dill seeds can be
   harvested by cutting off the flower heads when the seeds begin to ripen. Place these
   flower heads upside down in a paper bag and allow them to dry in a warm, dry place.
   After one week, remove the seeds and store in an airtight jar to use in flavoring foods.

Dill harvest grid:

Yield	Harvest Time	Tools	Harvest Readiness	Washing	Processing	Storage
.07 lb. per row foot	3–4 bins per hour	Knives and boxes	7–10 inches tall, dark green before it bolts	Dunk in cold water in two bins, repeat for 30 seconds in second bin	Remove dead and yellow leaves	Store in cool place with high humidity

Duck Potato

Usage: Water plants, edible roots and shoots

Species: Duck Potato (*Sagittaria latifolia*)

Growing: Also known as Broadleaf Arrowheads or Wapato, this plant produce a root vegetable that is eaten like potatoes. They grow directly in water or in marshy soil, and can be planted from eyes just like potatoes. When harvesting, no more than a quarter of them should be collected per year, and they will spread to replace the loss. Gather these in anytime during the summer and fall until the first frost.

Duckweed

Usage: Water plants, edible

Species: Common Duckweed (*Lemna minor),* Star Duckweed (*Lemna trisulca*), Swollen Duckweed (*Lemna gibba*)

Growing: This versatile plant looks less like a plant than a small green disk that just floats on the surface of water. It is very high in protein, containing even more than soybeans, and although it is commonly used to feed water livestock like fish and ducks, it is edible and eaten by people in some places in Asia. Once introduced to a freshwater environment, they spread quickly and cover the entire surface unless an animal lives there to eat them. On small ponds they help to prevent evaporation, and also provide shade to small fish.

Fennel

Usage: Pest control, edible

Species: Sweet Fennel (*Foeniculum vulgare*)

Growing: Fennel is easy to grow, and in fact, because of how quickly it spreads, is now on invasive species lists in North America. The dried seeds are used as a spice, and the root bulb and leaves are used raw or cooked. It has many medicinal uses. Be aware that Poison Hemlock looks very similar to fennel and grows near water or wet soil. Crush the leaves and smell them. Fennel has the distinctive smell of licorice, and Hemlock smells musty.

Fennel harvest grid:

Yield	Harvest Time	Tools	Harvest Readiness	Washing	Processing	Storage
1 bulb per row foot	75 bulbs per hour	Knives and bins	When bulb weighs ½ lb.	Do not wash	Free of blemishes and soft spots	Store in cool place with high humidity

Fenugreek

Usage: Edible, nitrogen fixing legume

Species: Fenugreek (*Trigonella corniculata*)

Growing: Fenugreek is an annual herb that enjoys full sun and well-drained soil. The seeds are the most useful part and can be harvested in early fall when the pods have dried. Store them in an airtight container. These are a useful herbal remedy and also used in Indian food for making pickles, curry, and sauces. The leaves can also be used in salads, although they have a particularly bitter taste.

Fig

Usage: Edible fruit, dryland tolerant

Species: Common Fig (*Ficus carica*)

Growing: Fig trees grow from twenty to thirty feet (6–9 m) tall and can grow in most places that experience a long, hot summer. They enjoy full sun and regular watering. Figs are high in calcium and one of the oldest and more nutritious human food crops. The fruit can be eaten fresh, dried, or as jam. Use caution when handling the tree, as the sap is a skin irritant. The tree produces two crops per year, one in the spring and one in late summer or fall. The first, or *breva*, crop is usually very small but a few varieties do produce a little more breva fruit.

Garlic

Usage: Pest control, edible

Species: Garlic (*Allium ampeloprasum*)

Growing: Garlic is part of the onion family and needs lots of nitrogen. It does well if it follows a crop of beans, and a mulch of hay is used to keep the weeds at bay and conserve moisture. After harvest, it should be cured in a greenhouse and then stored. The largest bulbs are saved for seeding. Scapes can be harvested when they are six inches long and have a small bulb.

Gooseberry

Usage: Bee forage, edible fruit

Species: Gooseberry (*Ribes uva-crispa*)

Growing: These berries enjoy partial sun to full shade and grow three to five feet (1–1.5 m) tall and wide. In some places, gooseberries are illegal, as they sometimes harbor a pest that kills white pines. They are very hardy and grow in most climates.

Garlic harvest grid:

Yield	Harvest Time	Tools	Harvest Readiness	Washing	Processing	Storage
0.2 lb. per row foot	75 lb. per hour	Boxes	First two leaves are starting to die back	Do not wash	Allow to cure and store when leaves crumble off necks	Store in cool dry place

Grape

Usage: Edible fruit

Species: Fox Grape (*Vitis labrusca*), Muscadine Grape (*Vitis rotundifolia*)

Growing: Unlike the varieties of grapes that are cultivated by humans, such as *vitis vinifera* and *vitis ripari*, Foxes and Muscadines are much hardier against pests and diseases. They are best grown on a trellis, which requires more effort in pruning but is worth the trouble. The trellis needs to be extremely strong because in a few years the grape will grow to be very heavy and treelike. Foxes are self-pollinating but Muscadines need several plants to make sure there are both male and female varieties. Foxes grow in any climate and Muscadines need the warm, humid climate found in the south. Grapes can be eaten raw or made into jam, juice, vinegar, wine, raisins, oils, or seed extracts. While grapes grow anywhere, they don't produce fruit unless they have a long, hot growing season, which is challenging in a northern climate. They enjoy full sun and the heat of the south side of a house, and they need regular watering in well-drained soil. It may not be worth growing them in cold climates, as the entire vine must be laid flat on the ground and mulched, and the buds must be protected from frost every year.

Groundnut

Usage: Edible roots and shoots, nitrogen fixing legume

Species: Groundnut (*Apios americana*), Fortune's Groundnut (*Apios fortune*), Price's Groundnut (*Apios priceana*)

Growing: A groundnut is somewhat like a "new" potato, of similar size, with lots of little roots coming off it, but unlike potatoes, groundnuts are full of protein. They climb and send out roots, invading the neighboring areas and so they must be kept in check. The roots can be cooked exactly like any other root vegetable.

Guava

Usage: Edible fruit

Species: Apple Guava (*Psidium guajava*)

Growing: While the guava is a subtropical plant, it is surprisingly hardy and can survive temperatures as low as 40°F (4°C) and on very little rain. As long as your climate does not experience frost and you protect young plants, you can grow guava. They can bear fruit even when grown in containers and usually do so within a couple of years. They are eaten raw or cooked, or made into juice. They do better with regular watering that reaches their deep roots, then allowing their roots to dry completely before watering again. They need fertile, well-drained soil and plenty of nitrogen in the form of manure. When the fruit is ripe, it will change in color and smell dramatically. They are best when allowed to ripen on the tree but green fruit can be stored for up to five weeks in cold storage.

Hawthorn

Usage: Edible fruit, bee forage, erosion control, windbreak, dryland tolerant

Species: Hawthorn (*Crataegus L.*)

Growing: There are many, many varieties of Hawthorn, which is a small tree growing around twenty feet high. It enjoys partial shade to full sun, and well-drained soil. The berries are edible and are usually cooked like apples, in jams and pies.

Hazelnut

Usage: Edible nut

Species: Common Hazel (*Corylus avellana*)

Growing: The hazel is a shrub that grows from ten to twenty feet (3–6 m) high. Traditionally, besides producing a protein-filled nut, it also serves the useful purposes of being part of a hedge and a fencing material. The flexible branches can also be made into plant supports and arches for climbing species. They prefer well-drained soil but do better in rainy areas, especially in the Pacific Northwest. Harvest the hazelnuts in midfall when the nuts and leaves begin to drop. Allow the nuts to dry, and then eat raw or roasted, or ground into a paste.

Hickory

Usage: Wood and timber, edible nuts

Species: Northern Pecan (*Carya illinoinensis*), Shellbark Hickory (*Carya laciniosa*), Shagbark Hickory (*Carya ovate*)

Growing: Hickories enjoy full sun and grow from 70 to 100 feet (21–31 m) tall. The Northern Pecan gets to be 75–120 feet (23–37 m) wide, while the hickories are 30–50 feet (9–15 m) wide. Hickories need other hickories nearby to pollinate, so you will need to plant a few. It is recommended to graft pecans onto other hickories. The nut will be improved and the pecans will bear sooner than the usual ten to fifteen years. The hickories and pecan species listed here combined cover all areas of North America, but you need to pick the right species for your climate. Soil types don't effect the hickory family much, but temperature does. The hickories are more valuable for their wood, which is used in smoking meat and makes excellent firewood. Hickory also makes very durable furniture. Hickory nuts are just as good as pecans and are all harvested as they start to fall to the ground. Shake off the rest. Hull them, wash the nuts, and spread them out to dry. The drying process in a well-ventilated room or garage with a fan takes about two weeks.

Honey Locust

Usage: Edible seedpod, dryland tolerant, nitrogen fixing

Species: Honey Locust (*Gleditsia triacanthos*)

Growing: The Honey Locust can reach a height of 60–100 feet (18–31 m) and lives to be about 120 years old. It is sometimes described as a perennial shrub rather than a tree. It prefers well-drained fertile soil. The pulp inside the seedpod is edible (although be careful as the Black Locust is not), and the pods make great forage for grazing animals. The tree is hardy, grows quickly, and provides massive areas of shade for people and animals.

Honeysuckle/Honeyberry

Usage: Wetland tolerant, edible

Species: Blueberried Honeysuckle (*Lonicera caerulea*)

Growing: There are many types of honeysuckles, but only a very few have edible berries. Russia has been raising Blueberried Honeysuckles as a food crop for a long time. It prefers wet and marshy soil, and is usually found in the Northeast. It has never been found wild in the Pacific Northwest. The fruit can be used for jam, juice, wine, ice cream, yogurt, and sauces. It can handle very cold temperatures and needs to have two compatible varieties in order to produce fruit. Harvest the fruit as it ripens.

Hops

Usage: Edible

Species: Common Hop (*Humulus lupus*)

Growing: Hops are a perennial climbing plant that produces flowers called hops that are used mostly in beer production. The extract from the hops is antimicrobial and is naturally sedative. Hops take up a lot of space and enjoy full sunlight. They need regular watering and well-drained soil that is rich in minerals and nitrogen. It takes about four months for them to produce flowers, but they grow well in most places as long as they are not allowed to freeze. They are susceptible to molds and insects, which must be prevented with a drip irrigation system and organic pesticides. The flowers are ready to harvest at the end of summer when the flower cone has become light and dry and doesn't stay compressed when squeezed. Dry these cones on a drying rack in the sun or in a dehydrator and stir and rotate them daily until the last cone is springy and the powder falls out easily. Seal them in a freezer bag and put them in a freezer until you are ready to use them.

Horseradish

Usage: Edible

Species: Horseradish (*Amoracia rusticana*)

Growing: Horseradish is related to cabbages and broccoli, and as such is a hardy and strong-flavored perennial that grows just about anywhere. It prefers a long growing season and a cold winter. The root is harvested after the first frost. A few pieces are cut off for replanting, while the rest is kept for yourself. If it is left in the ground, it will shoot out sprouts underground and take over the garden. Horseradish is a great flavoring for food but do not eat it raw in any great quantity, as it will cause digestive upsets. It must be grated and added to vinegar where it will keep for several months. When it begins to darken in color, it is time to throw it out. Horseradish sauce is simply the vinegar combination mixed with mayonnaise. It is also a valuable herbal medicine.

Hyssop

Usage: Bee forage, edible

Species: Herb Hyssop (*Hyssopus officinalis*)

Growing: Hyssop is a hardy perennial that stays short and can make a tiny hedge. It prefers full sun and well-drained soil. Hyssop has a slightly minty flavor and can be added to soups, but it is far more valuable as a medicinal herb with hundreds of uses. Harvest the plants just before the flowers begin to open, and hang upside down to dry. Remove the leaves and flowers and place them in an airtight container.

Jasmine

Usage: Wetland tolerant, edible, climbing vine

Species: Jasmine (*Jasminum L.*)

Growing: There are more than 200 different types of jasmine, and they grow all over the world. In Asia, they are grown for their flowers, which only open in the evening and are useful not only for their beauty, but as a flavoring for green tea. The flowers can also be used to make extracts of syrup and essential oil. There are two kinds of jasmine, one that stays green all year, and other species that lose their leaves in the winter. They enjoy warm, moist climates and regular watering.

Jerusalem Artichoke

Usage: Edible roots

Species: Jerusalem Artichoke (*Helianthus tuberosus*)

Growing: These produce an edible root vegetable that looks a little like a yam. Each plant produces tons of food and is likely to spread unless contained. They enjoy partial shade to full sun and can be grown in most climates. Jerusalem Artichokes are a perennial, not related to artichokes at all, and are actually closer to sunflowers. Because of the quantity of food produced by each plant, and the high sugar carbohydrate content, it is a good source of fructose. It can also be made into ethanol and yeast for fermentation. They are easy to grow, but every year you must dig them up and replant them in order to prevent them from taking over. They can be eaten cooked like potatoes or as a substitute for turnips and parsnips. These artichokes make excellent animal fodder, especially for pigs, which can dig them up themselves. They will cook much faster than potatoes and must be watched so they don't turn into mush.

Jicama

Usage: Edible roots and shoots

Species: Jicama (*Pachyrhizus erosus*)

Growing: Also known as Mexican turnips or Yam Beans, the Jicama (with the *J* pronounced as an *H*) has a tasty root resembling a potato and is very sweet. It is usually eaten raw and sometimes used in stir-fry. It can be prepared with a variety of spices and is often added to salads or dipped in salsa. However, be cautious: the rest of the plant is highly poisonous. Only the root can be eaten. It is not the most nutritious of foods but is full of fiber and has sugars that are beneficial to the friendly bacteria of the digestive system.

Jujube

Usage: Edible fruit, wood and timber, dryland tolerant

Species: Common Jujube (*Ziziphus zizyphus*)

Growing: Jujube grows in many climates and conditions and is very hardy, although it needs to have a long, hot summer and regular water to produce fruit. It can survive cold winters and extremely hot days in the summer. The fruit is edible and used in traditional medicine, as well as eaten raw or made into candied dried fruits and jams.

Kale

Usage: Edible leaves

Species: Kale (*Brassica oleracea*)

Growing: Kale should not follow other brassicas because it is in the same family, and just like all other brassicas, it needs lots of compost and nitrogen and does well after beans. Transplants should not be older than five weeks. They are susceptible to flea beetles, cabbage worms, and cabbage loopers, which can be prevented with a row cover. If they are grown after beans, no additional fertilizer is necessary.

Kale harvest grid:

Yield	Harvest Time	Tools	Harvest Readiness	Washing	Processing	Storage
1 bunch per row foot	120–150 bunches per hour	Knives and boxes	When 15–18 inches tall, bottom leaves dark green	Do not wash	Free of dead and yellow leaves	Store in cooler

Kiwifruit

Usage: Edible fruit

Species: Hardy Kiwifruit (*Aetinidia arguta*), Super-Hardy Kiwifruit (*Actinidia kolomikta*), Purple Hardy Kiwifruit (*Actinidia purpurea*)

Growing: These hardy kiwi species enjoy full sun but can be grown almost anywhere. Compared to the grocery store variety, they are much smaller and less fuzzy. They even taste sweeter. You will need to plant male and female trees for pollination and protect them from frost when they are young, but once established, they will grow prolifically and provide hundreds of pounds of fruit every year. They will grow very tall and the fruit may be difficult to reach unless you train it to a trellis, which requires much more effort in pruning but is worth it. The fruit can be stored like apples in cold storage.

Lavender

Usage: Bee forage, dryland tolerant

Species: English Lavender (*Lavendula angustifolia*)

Growing: Lavender enjoys full sun and well-drained or even sandy soil. If it is too moist or too well fertilized it can be susceptible to mold. The flowers are used to extract essential oil, which is a popular natural remedy. They are also edible—sometimes candied, added to teas, made into syrup, or dried. For most uses the flowers are harvested when they are open and at their brightest. Cut the stems when they are dry and cool, especially in the morning after the dew has dried. You can either hang the flowers upside down or remove the heads and spread them on a drying rack in the sun or in the dehydrator.

Leeks

Usage: Edible leaves and stalk

Species: Allium porrum (*Alliaceae*)

Growing: Leeks need lots of compost and nitrogen-rich fertilizer. If the leaves begin to yellow, a high-quality fish fertilizer works well. Don't overwater them or they will lose minerals and can get damaged. They can be transplanted when they are as close to the width of a pencil as possible, but don't wait too long after last frost.

Leeks harvest grid:

Yield	Harvest Time	Tools	Harvest Readiness	Washing	Processing	Storage
1 leek per row foot	75 bunches per hour	Knives, harvesting fork, boxes	1 inch in diameter, tall, dark green leaves	Dunk once and spray off roots	If free of blemishes and disease, cut off roots and trim top	Store in cool storage with high humidity

Lemongrass

Usage: Stream bank erosion control, edible

Species: Lemongrass (*Cymbopogon citratus*)

Growing: Lemongrass is a perennial herb that makes a grassy clump about three to five feet tall. It prefers full sun and rich, moist soil. You can fertilize monthly during the summer. Cats are its only real enemy because they eat the leaves. Lemongrass is a common ingredient in Asian cooking and is used in tea, soup, curry, and meats either fresh or powdered. Harvest the leaves any time after the plant is over a foot tall and dry, or use fresh.

Lespedeza

(see Clover)

Lettuce

Usage: Edible leaves

Species: Lettuce (*Lactuca sativa*)

Growing: Lettuce just needs a lot of compost to thrive. Transplants should not be older than three to four weeks. It can be planted in successions for maximum harvest, or harvested continuously as baby lettuce. It is susceptible to aphids and whitefly, which can be prevented with a floating row cover. If harvesting for mixed greens, a careful weeding schedule must be maintained. Harvesting is labor intensive but it can be more efficient if it is cut at the right spot. Cut above the dead and damaged leaves.

The Ultimate Guide to Urban Farming

Lettuce harvest grid:

Yield	Harvest Time	Tools	Harvest Readiness	Washing	Processing	Storage
1 head per row foot	160 heads per hour	Knives and bins	Whole heads should be firm, baby greens should be tender and healthy	Dunk whole heads in tub and remove dirt. Dunk mixed greens in two tubs for 15 seconds, then use salad spinner to remove water.	Free of dead and yellow leaves	Store in cooler

Leucaena

Usage: Stream bank erosion control, ground cover, wood and timber, edible, nitrogen-fixing legume

Species: Leadtree (*Leucana benth*)

Growing: The Leucaena species is rated as the most useful nitrogen-fixing legume. It is native to the American continents and there are twenty-four species of trees and shrubs from all different areas and climates, although it prefers warm climates and regular watering. It does not tolerate frost, although frost won't completely kill it. The seedlings must be protected from weeds and animals, but once established they are fairly hardy. The leaves are a green manure, the wood is an efficient firewood, it is an excellent animal forage and is a quick-growing erosion control plant.

Licorice

Usage: Nitrogen-fixing legume, edible

Species: Licorice (*Glycyrrhiza glabra*), American Licorice (*Glycyrrhiza lepidota*), Chinese Licorice (*Glycyrrhiza uralensis*)

Growing: Licorice is a perennial herb that enjoys full sun and well-drained soil. It takes a couple of years before it can be harvested. To harvest, the root is boiled to extract the licorice flavor and then evaporated to concentrate it. This can be made into a dry powder or syrup that is much sweeter than sucrose. It has many medicinal uses, but excessive use can cause liver and heart problems. Eating it too frequently can cause a rise in blood pressure.

Loganberry

Usage: Bee forage, edible

Species: Loganberry (*Rubus loganbaccus*)

Growing: Loganberries are hardy to pests, disease, and frost. They grow as a bramble rather than tall canes (like raspberries). Individual canes die after a couple of years and should be cut off. The fruit ripens early and produces for about two months. It should be picked when it has changed from a deep red to a deep purple color. They can be eaten raw or made into jams and dessert. Use them exactly like raspberries and blackberries.

Loquat

Usage: Edible fruit

Species: Loquat (*Eriobotrya japonica*)

Growing: The loquat tree is usually small, and stays around ten feet high. It prefers warm temperatures, full sun and well-drained soil. They are fairly hardy, doing well in any kind of soil but need regular watering and warmth. The loquat fruit is similar to apples and can be eaten raw or made into jam, jelly, and chutney. The leaves are edible and often used to make a nutritious and medicinal tea.

Lotus

Usage: Water plant, edible

Species: American Lotus (*Nelumbo lutea*), Sacred Lotus (*Nelumbo nucifera*)

Growing: There are many varieties of lotus plants, some more cold hardy than others, so double-check the species you have to determine its lowest temperature. They need full sun all day and are planted in water, which means that your pond cannot be shaded by any trees. They are usually planted as a tuber, or root bulb that is pushed into the soil in the shallow end of the pond and weighted down with a rock. They need lots of fertilizer, so they do well with fish, but large fish will eat them very quickly. These tubers are edible by humans as well and can be cut up and used in stir-fry.

Lucerne

(see Alfalfa)

Macadamia

Usage: Edible nut

Species: Macadamia Nut (*Macadamia integrifolia*), Prickly Macadamia Nut (*Macadamia tetraphylla*)

Growing: Macadamia trees are medium-sized evergreen trees that produce a delicious edible nut. It won't produce nuts until it is at least seven years old. It enjoys fertile, well-drained soil, warm temperatures, and regular watering. The roots are very shallow and the tree can be easily knocked down by strong winds. The nuts are harvested when they begin to fall off the tree, and must be gathered every week during the harvest season. Dry the nuts on a drying frame until the nuts rattle inside their shells. Once dry, they must be stored in an airtight container in cold storage or a freezer. They can also be roasted for extra flavor.

Maple

Usage: Edible pods, bee forage, wood and timber

Species: Sugar Maple (*Acer saccharum*), Box Elder (*Acer negundo*), Black Maple (*Acer nigrum*), Red Maple (*Acer rubrum*),

Growing: Maple prefers shade to full sun, and grows 75–100 feet (23–31 m) high with a crown 75–100 feet wide. It takes forty gallons (151 L) of sap to make one gallon (3.8 L) of syrup, so you will need at least a few trees, and while sugar maples produce more sap than other varieties, any species of maple can be tapped. They can be planted closely together and will grow in all but the sandiest soils. The one thing that does effect them is temperature. They prefer a temperate climate with a winter temperature of 0°–50°F. Beware: even though the sap is edible, the leaves are toxic, so it is better to keep animals away.

In March or April, when the temperature still freezes at night but thaws in the daytime, test to see if the sap is flowing by cutting a small gash in the side of the tree that has the most limbs. Some will be running and some won't. When a tree is running, drive in a maple syrup spout and hang a bucket. The sap will drip into the bucket from the spout. Some trees produce more and tastier sap than others, and some years are just bad years, so the whole process is out of your control. You will need to collect the sap in the morning and afternoon. Dump the sap into a very large pot or cauldron. Each tree will yield six to twelve gallons (23–45 L). Heat the syrup up to a low boil or simmer. You will have to stir it almost constantly to keep it from boiling over, so it is better to have more than one person doing this and take turns. As it boils down, keep adding more. The sap is done when its temperature is 219°F (103°C). Use a candy thermometer to measure this. Don't let it scorch or it will ruin the syrup. A gallon will weigh about eleven pounds (5 kg). Stop tapping your trees when the snow melts, the ground thaws, and the buds start to swell. The sap just won't taste good. You can make sugar with the sap by letting it harden and grinding it up into a powder.

Marigold

Usage: Pest control

Species: Mexican Marigold (*Tagetes cempasúhil*), African Marigolds (*Tagetes erecta*), French Marigold (*Tagetes patula*)

Growing: Marigolds enjoy full sun and most kinds of well-drained soil. Due to their pest control qualities, they are fairly hardy and don't need to be fertilized. They make excellent companion plants and are often used as a border plant, as they can even deter deer and rabbits.

Mayapple

Usage: Edible fruit

Species: Mayapple (*Podophyllum peltatum*)

Growing: Also known as American Mandrake, the Mayapple is a hardy perennial that enjoys well-drained soil and partial to full shade. Only the ripe fruit is safe to eat, and it does not ripen in May as its name suggests. Unripe fruit is toxic and causes severe digestive upset, and even ripe fruit can cause problems in large quantities. The seeds and peel are also not edible. However, once peeled and the seeds removed, the ripe fruit can be eaten raw or used in jam and pie.

Mesquite

Usage: Bee forage, dryland tolerant, nitrogen-fixing legume, edible, wood and timber
Species: Honey Mesquite (*Prosopis glandulosa*), Velvet Mesquite (*Prosopis velutina*), Creeping Mesquite (*Prosopis strombulifera*), Screwbean Mesquite (*Prosopis pubescens*),
Growing: Mesquite is found wild throughout the Southwestern US and even up into the Midwest. It is a tree that can reach twenty to thirty feet (6–9 m) high although the trees generally stay small enough to look like a shrub. It grows a very long taproot that has been known to reach down as much as sixty feet (18 m) to draw up groundwater, but also draws up surface water if it needs to. Therefore, it can handle very long droughts and is extremely hardy. It grows and spreads quickly, and for this reason some people consider it a pest. The bean pods can be made into jam or dried and ground into delicious flour for bread. To harvest the pods, wait until they are beginning to fall from the tree. If they take more than a slight pull to come off the branch, they aren't ready yet. They should be as dry as possible. Rinse them, and then lay them out in the sun on a drying rack until they snap in two when bent. To make flour, the whole pod is ground, but this would gum up a regular grain mill. This can be done with a heavy-duty blender or a hammer mill. The meal or flour can be substituted for no more than half a cup per cup of grain flour. It is very nutritious. The pods also serve as animal forage, and the wood is hard and useful for furniture and other well-used items. The Honey Mesquite in particular is also used for decorative woodworking. Mesquite is commonly used as firewood and especially in barbecuing to add flavor to meat.

Mint

Usage: Wetland tolerant, bee forage, ground cover, edible
Species: Apple Mint (*Mentha suaveolens*), Bowles's Mint (*Mentha x villosa*), Field Mint (*Mentha arvensis*), Corsican Mint (*Mentha requienii*), Peppermint (*Mentha piperita*)
Growing: This nice-smelling perennial herb enjoys partial shade and quickly spreads everywhere. They tend to prefer moist soil, although they will grow anywhere. Due to their spreading capacity, they are often grown in containers to prevent their takeover of the garden. However, Apple and Bowles's make great ground covers and they do repel pests. The leaves are very healthy, and can be used raw, dried, or in teas. These can be harvested throughout the season.

Moringa

Usage: Edible

Species: Moringa (*Moringa oleifera*)

Growing: Moringa has huge potential to provide highly nutritious food. It prefers warm
temperatures and well-drained soil, and it needs full sun. It requires regular watering.
The tree grows very quickly, can reach fifteen feet (5 m) in one year, and has the capacity
to be cut down every year to three feet (1 m). It will grow back and will produce edible
pods that are within easy reach for picking. In several years it will produce thousands of
pods. The leaves and pods can be used as animal forage without any processing, and
they can also be harvested for human consumption. The leaves are used raw in salads,
dried and used as tea, or ground into powder and added to food for its intense nutritional
value. The flowers can be harvested and must be cooked and used like mushrooms. The
pods can be used fresh raw or cooked in stir-fry and other dishes, or the seeds can be
removed and cooked like peas.

Mulberry

Usage: Edible fruit, dryland tolerant

Species: Red Mulberry (*Morus rubra*), White Mulberry (*Morus alba*)

Growing: Mulberries grow in most climates and will still produce fruit in partial shade.
While many are somewhat tasteless, a variety can be chosen that has a stronger flavor,
making them a very low-maintenance food provider. The real value of mulberry,
although it is great raw or as pie, are the nutritional and herbal medicinal uses, which
are very high. The Mulberry leaves are also the sole source of food for silkworms, if
spinning silk is of interest to you. Silk spinning is actually quite easy because the fibers
are already long. It is very durable and lightweight.

Myrtle

Usage: Edible fruit

Species: Common Myrtle (*Myrtus communis*)

Growing: Myrtle is an evergreen shrub that is often used in topiaries. They can't handle
extremely cold temperatures, but if they are planted near shelter, they will be fairly cold
hardy. They prefer well-drained soil. The leaves have been used as herbal medicine for
thousands of years, and the berries are edible, although they are very bitter.

Nasturtium

Usage: Ground cover, edible

Species: Any varieties

Growing: Nasturtiums are easy to grow and enjoy partial shade to full sun. The entire plant is
edible, but people love to use the flowers since they are so brightly colored. They repel
pests and are generally hardy. The unripe seedpods are often pickled to make a peppery
condiment, or the ripe seeds can be dried and used exactly like peppercorns to replace
pepper. The flowers and leaves are used raw or cooked to add a peppery flavor to any
dish.

Natal Plum

Usage: Edible fruit, ground cover

Species: Natal Plum (*Carissa macrocarpa*)

Growing: The Natal plum does not produce plums, but is a shrub that produces a berry that is shaped like a plum. Although tart, it is very nutritious. The plant can also be used as a hedge because of its thorny branches without much care or effort, as it doesn't need much pruning, and dwarf varieties can provide a very effective ground cover. The plant grows in most conditions and doesn't even really need regular watering. It takes a few years before it will produce fruit. Pick the fruit when ripe and be careful not to bruise it. You can eat the berries raw but they taste better as a substitute for cranberries or in jam.

Narrow-leaved Tea-tree

Usage: Pest control

Species: Narrow-leaved Tea-tree (*Melaleuca alternifolia*)

Growing: This plant produces the highly popular Tea Tree Oil that is so heavily marketed today. They are native to the hot weather of Australia but have been grown with varying success in many other places. They enjoy full sun and only occasional watering, and must be protected from frost when young. They aren't an especially useful plant in the garden, but are included here (like Neem), for the vast range of useful products that can be made from the oil. The leaves are picked and the oil is distilled and bottled. It is antibacterial, antifungal, antiseptic, and antiviral. It is not edible, but it can be used for just about any skin problem. It is an extremely effective pest control.

Neem

Usage: Pest control, dryland tolerant

Species: Neem (*Azadirachta indica*)

Growing: Neem is a subtropical evergreen tree that grows quickly. It is tolerant to drought and prefers well-drained soil. None of it is edible and it has very little use in the garden except as a soil input, but it is included for the vast array of products that can be made from it. The entire tree has antifungal, antibacterial, sedative, and antiviral properties. It is commonly used as a natural pesticide. The extracts have been used to make toothpaste, skin creams, sprays, and a plethora of other items to cure just about every type of skin ailment.

Nettle

Usage: Edible, groundcover

Species: Stinging Nettle (*Urtica dioica*)

Growing: Nettles grow as either annuals or perennials and have stinging hairs on the leaves and stalks that greatly irritate the skin. Despite this, they are a very old medicinal plant that can be used for teas, shampoos, and skin salves. They are very hardy and grow similarly to blackberries, which happen to be their direct competitor. Nettles are high in vitamins and minerals and it only takes soaking or cooking to remove the stinging. It is often used in polenta, pesto, soup, and cooked greens. It is also a substitute for flax or hemp in making linen and the processing is the same. It deters pests and if kept under control is a great companion plant. Not all stinging varieties of nettles are edible, so make sure you are eating the right one. To harvest the leaves, wear a pair of gloves and use a pair of scissors to cut the leaves off in March or April before they begin to flower. The smaller leaves near the top are the best. Soak them in warm water for ten minutes, and drain off the water without touching it. They are now ready to cook.

New Zealand Spinach

Usage: Dryland tolerant, edible, ground cover
Species: New Zealand Spinach (*Tetragonia tetraconioides*)
Growing: This leafy vegetable is very similar to spinach and is cooked like spinach. It prefers a hot, moist environment. It spreads rapidly so it is considered invasive in some places, but this quality also makes it an excellent ground cover. It has no enemies and even slugs don't like it.

Olive

Usage: Edible fruit, dryland tolerant
Species: Mission (*Olea L.*)
Growing: Olive trees enjoy the limestone soils near the ocean and prefer sandy and clay soil. They don't do well in fertile ground. They need hot weather and can't tolerate temperatures below 14°F (−10°C), but they are also very hardy to drought. The trees are susceptible to pests and fungus, which can be prevented with predatory wasps, pruning, and by avoiding fertilizers. The olives are harvested in late fall when they begin to fall off the tree. For making oil, the olives do not need to be handled gently, but for other uses they must be picked by hand. Mission olives do not need to be fermented before processing like green olives, but they do need to be cured. Wash the olives, allow them to dry thoroughly and put them into a well-ventilated box lined with burlap or other cloth. Vegetable crates work well for this purpose. Mix in one pound (0.5 kg) of salt per two pounds (1 kg) of olives, and cover it all with another inch (2.54 cm) of salt. This box should be placed where it can drain, either on a waterproof tray or onto the ground. Leave them there for one week. Starting the second week you will have to mix them thoroughly every three days. Pouring them into another container and back again is an easy way to do this and it's a good opportunity to pick out the soft or broken ones. Continue doing this every three days for three weeks, which may seem tedious, but it's worth it. By this time, a month has now passed and the olives should be shriveled up. Use a strainer to remove the salt and dip the olives for a few seconds into boiling water. Allow them to dry again completely and mix them with salt, this time one pound of salt for every ten pounds (4.5 kg) of olives. Put them into an airtight container and store in a cool place. In the cold storage, they will last a month, but in the fridge or freezer they will last longer.

Onion

Usage: Edible roots and shoots

Species: Multiplier Onions (*Allium cepa aggregatum*), Egyptian Walking Onion (*Allium cepa proliferum*), Nodding Wild Onion (*Allium cernuum*), Welsh Onion (*Allium fistulosum*), Chives (*Allium schoenoprasum*), Garlic Chives (*Allium tuberosum*)

Growing: All the species above are perennials, which means that you don't need to replant every year. You can eat the entire plant, but to keep them going the next year, simply divide the clump in half and leave the rest in the ground. They grow incredibly easily, are cold and pest hardy, and will grow just about anywhere. The greens (or green onions) can be cut throughout the season, and the bulb is also harvested in the fall.

Onion harvest grid:

Yield	Harvest Time	Tools	Harvest Readiness	Washing	Processing	Storage
1 green onion bunch per foot, 2 lb. onions per row foot	40–60 green onion bunches per hour, 175 lb. onions per hour	Knives and bins	Green onions are white and 3–4 inches tall	Do not wash, just spray dirt off roots	Free of dead and yellow leaves; cut off tops of storage onions 2–4 inches from crown and cure in greenhouse until neck is dry	Store in cooler

Palm

Usage: Edible, wood and timber

Species: Coconut Palm (*Cocos nucifera*), Jucara Palm (*Euterpe edulis*), Açaí Palm (*Euterpe oleracea*), Peach Palm (*Bactris gasipaeas*), Cabbage Palm (*Sabal palmetto*)

Growing: There are hundreds of types of palms living in almost every climate, and yet many are endangered due to their usefulness as a source of timber, which is made worse by their slow growth. Palms require partial shade to full sun, deep watering just after planting and then no water unless they are extremely dry. The leaves gradually turn brown and fall off, but you can cut those off to keep them pretty. They will grow in even cold climates—as long as they are somewhat insulated in the winter. The Cabbage Palm is cold, fire, wind, and drought resistant. The Peach Palm is more of a warm-climate tree, and it produces an edible fruit that can be eaten raw or cooked. It is often made into jams or flour. The Açaí Palm also produces a fruit that is smaller than a grape that is extremely popular in Brazil and served with everything because of its amazing

The Ultimate Guide to Urban Farming

nutritional value. The leaves can be made into hats, baskets, and roof thatch, and the wood can be made into furniture.

The Jucara is almost exclusively used for heart of palm, but all of these varieties can be harvested for heart of palm. Heart of palm is a delicacy, but for species that grow fruit, it does not make sense to cut down the whole tree to eat its heart. However, should you find yourself with a downed palm, you should harvest the valuable heart. To do this, you must peel off the outer layers to reveal the white central core. This is very labor intensive. It should be canned, fermented, or eaten fresh. The other edible product of palms is, of course, the nut of the Coconut Palm. The brown coconut you see in movies is actually only a small part of the coconut, which has a large husk around it. When they are ripe the husk will be a bright green color and must be cut off with a machete. The coconut can be processed into coconut oil for cooking, the meat can be eaten fresh or dried, the milk can be used raw or condensed, and the flour can be used in baking. Even the husk is useful in making charcoal, and the shell is used for cups and bowls all over the world. Beware: The coconut is a common allergen and can cause food and contact allergies.

Palo Verde

Usage: Seedpod, dryland tolerant, nitrogen-fixing legume
Species: Blue Palo Verde (*Parkinsonia florida*), Yellow Palo Verde (*Parkinsonia aculeata*)
Growing: Palo Verde is a small tree with a green trunk that is difficult to grow but once
 established, it is very hardy. They grow in dry, hot places but do need regular watering.
 It provides a shady canopy for other plants. Harvest the pods when they turn green,
 or let them dry on the tree. They are dry when they turn completely brown. Wash
 the green pods and blanch in boiling water for a minute and a half. Dunk them
 immediately in ice water for another minute and a half and store them in the freezer.
 Dry pods must be laid out to dry a little longer; then walk on them to crush the pods.
 Winnow away the shells and store the seeds in airtight containers. The flowers can also
 be eaten raw or candied.

Parsnip

Usage: Edible
Species: Parsnip (*Pastinaca sativa*)
Growing: Parsnip is a cold-climate crop that needs frost to fully develop. They can grow in
 any type of soil except the rockiest. The root is the edible part, but the leaves irritate
 the skin. Handling the plant during harvest requires gloves to prevent burning and
 blistering. The root is even more nutritious than carrots and can be eaten raw or
 cooked. It is commonly boiled, roasted, or used in soups. They will be ready to harvest
 in mid-fall and taste much better after the first frost. Dig up the root and store in cold
 storage.

Parsnip harvest grid:

Yield	Harvest Time	Tools	Harvest Readiness	Washing	Processing	Storage
1 pound per row foot	175 lb. per hour	Knives and bins	Roots 6–12 inches long	Do not wash until delivered then hose them down	Twist off tops, remove any stunted and diseased parsnips, or ones with splits	Store in cool area with high humidity; small parsnips do not store well

Passion Fruit

Usage: Edible fruit, climbing vine

Species: Passion Fruit (*Passiflora edulis*)

Growing: There are two types of passion fruit: the yellow variety and the purple variety. The yellow kind is very big and grows fruit up to the size of a grapefruit, and the purple kind often tastes better but stays smaller than a lemon. They don't tolerate frost but are very cold hardy and can withstand almost freezing temperatures. Even if it does frost, a mature plant is so bushy that it insulates itself and comes back. It needs a strong trellis to support the fast-growing vine. They enjoy partial shade to full sun and well-drained fertile soil. Snails and a variety of diseases plague them, but with proper prevention a good crop is still possible. The fruit will turn ripe over a period of only a few days and quickly turn purple or yellow and fall off. Gather them from the tree and ground, wash and dry them without injuring them, and store them in cold storage. They will last up to three weeks. Freeze or juice to preserve.

Paulownia

Usage: Wood and timber, edible, bare soil erosion, pest control, bee forage

Species: Paulownia (*Paulownia fortunei*)

Growing: The paulownia is a very fast-growing hardwood tree that can provide timber in as little as five years. They are touted as a solution to the world's unsustainable tropical wood industry. The leaves make excellent animal forage with high protein content, and they can be intercropped easily with grains and other plants. They need full sun and deep regular watering when the soil is dry. Additional fertilizer can be added throughout the year. In the first year, the tree should grow to at least ten feet tall. If it does not, cut it down in early spring before the leaves are formed. This will make it grow twice as fast the next year. The tree can be cut back like this almost every year. The trees are only really susceptible to frost when young, but otherwise can be grown in most climates, except for the very coldest places.

The Ultimate Guide to Urban Farming

Paw Paw

Usage: Edible fruit

Species: Pawpaw (*Asimina triloba*)

Growing: Some people inaccurately call this plant a papaya but the Paw Paw is not the same as the tropical fruit and is native to North America. It enjoys partial shade to full sun and is a pest-resistant and generally hardy plant. It requires warm, humid temperatures and provides large, delicious fruit. The ease of growing and prolific supply of fruit makes it highly recommended for the home orchard in the right climate. It grows twenty to thirty-five feet (6–10 m) tall and twenty to thirty-five feet wide and needs fertile, well-drained soil. Be aware that the seeds, leaves, and wood are toxic and sometimes made into pesticides. To harvest, wait until the fruit is just a little bit soft and has changed from green to yellow or brown. It will keep for about three weeks in the fridge. The fruit can be eaten raw or made into pie, bread, pudding, and jam, and can be used in place of bananas.

Pea

Usage: Nitrogen-fixing legume, edible, climbing vines

Species: Butterfly Pea (*Clitoria mariana*), Beach Pea (*Lathyrus japonicas maritime*), Everlasting Pea (*Lathyrus latifolius*), Earthnut Pea (*Lathyrus tuberosus*), Carolina Bush Pea (*Thermopsis villosa*)

Growing: Peas are so easy to grow. They enjoy full sun and well-drained fertile soil, and they need trellising to keep them up. The peas will be ready to harvest in late summer and can be picked continuously until frost. The more you pick, the more will grow. The leaves of some of these peas are toxic. Some, like Snow Pea, are eaten fresh with the entire pod. Others are picked, shelled, and the peas dried.

Pea harvest grid:

Yield	Harvest Time	Tools	Harvest Readiness	Washing	Processing	Storage
¾ pint per row foot	20 lb. per hour	Bins/ buckets	Snow/snap peas should be flat and crisp, at least 2 inches and not filled out; other peas should be filled out	Do not wash; dunk in ice water to remove field heat	Remove diseased, damaged, and overmature pods	Store in cooler

Peach

Usage: Bee forage, edible

Species: Peach (*Prunus persica*), Nectarine (*Prunus persica var. nectarine*)

Growing: Peaches and nectarines need cold winters and long, hot summers. At the same time, the spring flowers can't tolerate frost, which prevents a fruit crop. This makes them an excellent temperate climate crop. They enjoy full sun, good airflow, regular watering, and frequent nitrogen fertilizer. When the fruit grows to almost an inch wide, it is common to thin them out so that the rest of the fruit will be sweeter. Pick them on a hot day when they are just a little soft and have a sweet smell. They must be eaten or preserved immediately as they can't be refrigerated or stored in live storage.

Peanut

Usage: Nitrogen-fixing legume, edible

Species: Peanut (*Arachis hypogaea*)

Growing: The peanut grows best in sandy soil. It needs a long growing season and regular watering. The pods must harvest at just the right time, when they are fully ripe but before they detach into the ground. When this happens the leaves will begin to turn yellow and the peanut skin inside the pod will be like paper and a light pink color. To harvest, dig up the entire plant gently and shake off the dirt. Then flip it upside down in a dry, shady place for a few days until the pods are mostly dry. Thresh and remove the pods from the plant. Peanuts may be eaten raw or roasted, ground into butter, or the oil extracted for use in cooking. Unshelled nuts will last for nine months in the fridge, and in the freezer will last forever if blanched for three minutes and cooled in ice water for three minutes. To make peanut butter, roast the nuts for 20 minutes at 300°F (149°C), stirring now and then, and remove the shells. Process in a blender with half a teaspoon of salt per cup of peanut butter.

Pear

Usage: Bee forage, edible fruit

Species: Asian Pear (*Pyrus bretschneideris*), European Pear (*Pyrus communis*)

Growing: Pears enjoy full sun. The Asian variety grows quite a bit taller than the European variety, but neither gets wider than twenty-five feet (8 m). They are much hardier and more pest-resistant than apples, with the exception of fire blight, which you can prevent by getting a resistant variety. Pears come in dwarf varieties and there is a plethora of species to choose from. You will need two trees for pollination, but they can be from either species. Pears are somewhat unique in that you harvest them before they are ripe, with the only clue being that the stem will begin to detach. The pear will start to hang at an angle rather than vertically. The one thing apples have on them is that pears aren't as easy to store and preserve, although they are stored similarly. They need to be chilled, usually in a cold storage, in order to finish ripening, where they will stay good for a few months. They shouldn't be stored with onions, cabbage, carrots, or potatoes, or they will absorb the smells. They can also be made into jams and sauces.

Pecan

(see Hickory)

Persimmon

Usage: Edible fruit

Species: American Persimmon (*Diospyros virginiana*)

Growing: The persimmon is the apple of Asia and is one of the most popular fruits there. In North America it hasn't caught on, possibly because it tastes awful right until it reaches ripeness, at which point it becomes sweet and delicious. It prefers warmer, humid climates and enjoys full sun. The size of the tree really varies, and you will need one male persimmon for every eight female trees, for pollination. There are hybrid varieties that grow larger fruit and come in dwarf sizes that may lend themselves to urban settings.

Pigeon Pea

Usage: Wood and timber, edible, windbreak, nitrogen-fixing legume, dryland tolerant

Species: Pigeon Pea (*Cajanus cajan*)

Growing: Pigeon Peas are a very ancient source of food and grown all over the world. They can grow in most soil types and are resistant to drought, making them extremely hardy and versatile. They can be harvested as a pea and eaten raw, dried, or ground into flour. Their nutritional value is very high, and when mixed with a cereal crop, they make a very balanced source of protein. They are especially healthy when sprouted and cooked. They are also a very useful and easy-to-grow animal forage.

Plum

Usage: Bee forage, windbreak, edible fruit

Species: American Plum (*Prunus Americana*), Canada Plum (*Prunus Americana* var. *nigra*), Chickasaw Plum (*Prunus angustifolia*), Hog Plum (*Prunus hortulana*), Beach Plum (*Prunus maritima*), Wild Goose Plum (*Prunus munsoniana*)

Growing: Plums enjoy full sun, and there is a species for most climates. They are susceptible to pests and diseases, and they need several to pollinate. A small stand of plum trees can be very tasty if the right species and tree is chosen, and hopefully will not need too much care to protect them. To harvest the plums, usually the first sign of ripening is a change in color, although that is still not a hard indication of readiness. They will also feel slightly soft and the skin will acquire a powdery texture. The plums must be eaten immediately or processed into jam right away.

Pomegranate

Usage: Dryland tolerant

Species: Pomegranate (*Punica granatum*)

Growing: There are many different varieties of pomegranate, but they are all very similar in the way that they grow. The differences lie mainly in the intended use and the color of the fruit. They are very drought resistant and can tolerate freezing—although they need

---

Plants

a long growing season to produce fruit. They enjoy partial shade to full sun. The fruit is ready to be harvested when it makes a metallic sound when tapped. If they aren't picked they will crack open and be ruined. They can't be picked, but have to be clipped off close to the fruit without leaving a stem. They can be stored like apples in cold storage for seven months as long as the temperature remains steady and the humidity does not rise too high. They are eaten by cutting them open and removing the juice sacs, or you can simply squeeze and juice them.

Poplar

Usage: Stream bank erosion control, wetland tolerant, wood and timber

Species: Eastern Cottonwood (*Populus deltoids*), Fremont Cottonwood (*Populus fremonti*), Black Poplar (*Populus nigra*), White Poplar (*Populus alba*), Narrowleaf Cottonwood (*Populus angustifolia*)

Growing: Poplars and Cottonwoods grow in most climates and even subarctic conditions, and they grow very quickly. Be careful not to plant them near any underground wires or pipes, as their extensive root system spreads up to 150 feet (46 m) away from the tree. They enjoy fertile soil and regular watering, partial shade to full sun. They are generally very simple to grow.

Potato

Usage: Edible roots

Species: Potato (*Solanum tuberosum*)

Growing: Potatoes ripen in cool weather and stop when the temperature gets too hot, so they do better in northern areas. However, while they can withstand a light frost, they must be harvested before a heavy frost. They are planted from the *eyes* of other potatoes, or the indented brown spots. During growth, the tubers tend to slowly pop out of the ground, but exposure to sunlight creates a green toxic spot that is inedible even with cooking; these potatoes must be covered up with more soil. Once they ripen, the whole plant is dug up carefully and cured. Curing requires keeping them at 65°F (18°C) for ten days in a well-ventilated crate in a place that is very humid. Once cured, potatoes can be put in live storage or dried. Beware: the rest of the plant is highly toxic, including any fruit that it might rarely produce above ground. It is a member of the nightshade family and although it looks like a tomato plant, it is deadly. The potatoes can only be eaten cooked and green parts must not be eaten.

Prickly Pear

Usage: Edible, dryland tolerant

Species: Prickly Pear (*Opuntia ficus-indica*) Eastern Prickly Pear (*Opuntia compressa*)

Growing: Prickly Pears grow prolifically in the Sonoran desert and produce a beautiful, edible fruit. Although it is a cactus, it is very cold hardy and can grow in northern areas as well. It enjoys full sun and well-drained sandy soil. If it gets too much water, it will rot and collapse, so drainage is key. The entire plant is actually edible, as long as the cactus spines and seeds are removed. To harvest the *nopales*, or leaves, remove one of the pads when they are about the size of your hand. This is done in spring and summer and you must wear gloves. Pick off the spines and use a vegetable peeler or paring knife to remove the skin and eyes. This must be done carefully and thoroughly. They are highly nutritious and can be eaten raw or cooked and used like green beans or a topping on Mexican dishes. To harvest the fruit, pick the smooth, firm, and shiny fruits, which will ripen for about one week in late fall or early winter. Wear gloves to do this. Use pliers to remove the spines, cut off the ends, and use a paring knife to remove all the skin layers from top to bottom. Be very careful to remove every part of the skin, as it has tiny hairs that will hurt your insides. Remove the seeds. These can be eaten raw or cooked, or made into jam or juice.

Quinoa

Usage: Edible seeds, dryland tolerant

Species: Quinoa (*Chenopodium quinoa*)

Growing: Quinoa is a hardy, protein-packed food source that can be grown in cooler
climates. Not only are the leaves some of the most nutritious greens, the seeds make
an excellent grain. Quinoa enjoys well-drained soil, is dryland tolerant, and prefers
climates that don't get warmer than 90°F (32°C). You don't need to water until they
have two or three leaves, and they can grow with very little water. You can harvest
some of the leaf greens when they are young and use them for salads. To harvest the
seeds, wait until the leaves have fallen or even until right after the first frost. The only
requirement is that the seeds must be very dry. If it rains, the seeds could germinate,
so the timing must be right. You should barely be able to make a dent with your
thumbnail in the seed. To remove any debris, place the quinoa on a screen and rub
it, and then the chaff can be blown away with a fan. Quinoa must also be rinsed to
remove the *saponin*, which is a bitter substance. Put it in a blender at the lowest speed,
and blend until it gets too frothy and bubbly to continue. Keep changing the water until
it no longer gets frothy. Alternatively, you could put it in a pillowcase and run it in the
cold water cycle of your washing machine. Cook it the same as you would rice.

Radish

Usage: Edible roots and shoots

Species: Radish (*Raphanus sativus*)

Growing: Radishes enjoy full sun and well-drained soil. There are many varieties and they
can be grown in most places. Some ripen in a month and can be replanted for several
harvests throughout the season. Their long taproot breaks up hard soil, although it stunts
their growth. Some radishes are harvested when small, and some grow to be the size of
a potato. The entire plant is edible, and some are harvested for their spicy seeds and the
greens used in salads.

Raspberry

Usage: Bee forage, edible fruit

Species: Red Raspberry (*Rubus idaeus*), Black Raspberry (*Rubus occidentalis*), Creeping
Raspberry (*Rubus tricolor*)

Growing: Everyone has eaten a raspberry, because there are hundreds of species everywhere. They self-pollinate and make better forest garden companions because they are less likely to take over than a blackberry. Trellising helps to keep them under control, unless you want to use Creeping Raspberry as an effective ground cover. They enjoy full sun and well-drained, fertile soil. They are easy to grow and are considered a little invasive because they spread so quickly. They are a little susceptible to fungus and shouldn't be planted where tomatoes, potatoes, peppers, eggplants, or bulbs have been grown before. The berries are harvested when they change to their deepest hue and can be easily pulled from the branch. They come in gold, purple, black, or red so be aware of the species you have. These berries are eaten raw, frozen, or made into jam. The leaves are an effective herbal medicine. Raspberries are extremely nutritious and high in fiber.

Radish harvest grid:

Yield	Harvest Time	Tools	Harvest Readiness	Washing	Processing	Storage
½ bunch per row foot	100 bunches per hour	Boxes	¾ inch–1 inch in diameter	Soak 25 bunches while you spray 2–4 bunches at a time	Free of dead and yellow leaves	Store in cooler

Rhubarb

Usage: Edible stalks, pest control

Species: Himalayan Rhubarb (*Rheum austral*), Rhubarb (*Rheum x cultorum*), Turkey Rhubarb (*Rheum palatum*)

Growing: Most gardeners grow rhubarb, which often persistently comes up in the compost pile. The leaves and roots are poisonous, but the stalks are delicious in pie (especially with strawberries). It needs full sun and rich soil. The stalks can be harvested throughout the season. Only choose the firm, medium-sized stalks. Old stalks are too firm for cooking. Cut off the leaves and roots and discard in the compost pile.

Rosemary

Usage: Bee forage, dryland tolerant

Species: Rosemary (*Rosmarinus officinalis*)

Growing: Rosemary is very easy to grow and very pest resistant and drought hardy. It needs full sun and well-drained soil. It has often been used in making topiaries. The leaves can be harvested throughout the season as a culinary herb that is high in iron and calcium. It is also a popular medicinal herb, although it is toxic in high doses.

Russian Olive

Usage: Edible fruit, nitrogen fixing legume, windbreak

Species: Russian Olive (*Elaeagnus angustifolia*)

Growing: Russian olives aren't related to olives, although they look similar. It is considered
    an invasive species in many places because it spreads so quickly and does well even
    in the some of the worst soils. They produce fruit in as little as three years. The fruit is
    edible although not particularly remarkable in taste.

Sage

Usage: Bee forage

Species: Common Sage (*Salvia officinalis*)

Growing: Sage is a common, easy-to-grow perennial herb that is used for its leaves in
    cooking and herbal medicine. The essential oil is commonly extracted for medicinal
    purposes. They enjoy partial shade to full sun and well-drained soil. It takes a year for
    sage to get established, but it will continue to provide leaves for at least a few years.

Salsify

Usage: Edible roots and shoots

Species: Salsify (*Tragopogon pornifolius*)

Growing: Salsify was once a more popular plant for vegetable gardens in North America
    and produces an edible root that tastes like oysters. The plant needs regular watering
    and loose soil. It is dug up in the fall to be eaten, or you can leave it in the ground over
    the winter to let it sprout again in the spring. The root must be peeled and washed and
    put in water with lemon juice or vinegar to prevent oxidation. Then it can be boiled and
    mashed, fried, steamed, or used in soups or stew.

Sloe

Usage: Bee forage, edible, wood and timber

Species: Sloe (*Prunus spinosa*)

Growing: Sloe, also known as blackthorn, is a large shrub that grows up to fifteen feet (5 m)
    tall and produces a purple fruit. These look much like Cherry plums, but the flowers
    are cream rather than white and bloom later. Sloe has large thorns, which make it
    a great hedge plant, and the wood is useful for firewood or carpentry projects. The
    fruit can be used in juice, wine, or jam, and some recommend that these should be
    harvested after the first frost for better flavor.

Sorghum

Usage: Edible, bee forage, dryland tolerant

Species: Sorghum (*Sorghum bicolor*)

Growing: Sorghum is very hardy and tolerates drought and high temperatures. It can be
    used as a grain, processed into molasses, and used as animal fodder. They need a
    long growing season with warm weather and lots of fertilizer. As a grain it is raised and
    harvested the same as wheat or corn, although you might need to wait until the first
    frost to allow it to dry enough. The plants cannot be used as animal fodder until they
    are at least eighteen inches (0.5 m) tall as the young shoots are poisonous. To make
    your own molasses, chop them down at the ground and remove the leaves. The stalks,

or *canes*, must be pressed or ground and the juice is caught in a container. This juice is then boiled down like maple syrup, until it is highly concentrated and sweet. This must be stirred continuously to prevent scorching. The syrup can be stored for many months, although it will begin to harden at the bottom. This hard sugar can be used as a sweetener or candy as well.

Spinach

Usage: Edible leaves

Species: Spinach (*Spinacia oleracea*)

Growing: Spinach needs lots of compost but doesn't really need a nitrogen-rich fertilizer. The seedbed needs to be weed-free and spinach can be succession planted by following peas, with lettuce after.

Spinach harvest grid:

Yield	Harvest Time	Tools	Harvest Readiness	Washing	Processing	Storage
½ lb. per row foot	5 bins per hour	Knives and boxes	3–4 inches tall, dark green, free of mold or yellowing	Dunk in water to wash, then dunk in ice water to cool, and rinse again for no more than 15 seconds	Free of dead and yellow leaves	Store in cooler

Squash

Uses: Edible fruit

Species: Zucchini, butternut squash, etc. (*Cucurbita pepo*)

Growing: Squash cannot follow other curcubits or nightshades such as cucumber or tomatoes, but it does well after brassicas or beans. It needs lots of compost and nitrogen-rich fertilizer. It grows very fast so make sure to transplant before it gets leggy or it can suffer in the heat. Usually black plastic mulch is used to warm up the soil temperature with composted chicken manure, and floating row covers prevent cucumber beetles. They are also susceptible to squash bugs and mildews. If they get squash bug larvae all you can do is destroy that crop. You can prevent squash bugs by removing the plastic and planting a cover crop right after harvest.

Stone Pine

Usage: Edible nut, dryland tolerant

Species: Stone Pine (*Pinus pinea*)

Growing: This tree can grow sixty feet (18 m) tall or more and produces one of the most ancient foods: the pine nut. It is a beautiful tree with its characteristic umbrella shape growing high above the tall stalk. They are very hardy, tolerant to drought, and take a long time to grow. They enjoy full sun and well-drained soil. When it does finally produce cones, the seeds are edible and often used in Italian pasta sauces.

Squash harvest grid:

Yield	Harvest Time	Tools	Harvest Readiness	Washing	Processing	Storage
4 lb. per row foot	120 lb. per hour	Knives and bins	Squash is 4–5 inches long and thick. Patty pans should be doorknob size. Firm all over.	Do not wash	Free of soft spots and squash bug larvae	Store in barn

Strawberry

Usage: Edible fruit, ground cover

Species: Garden Strawberry (*Fragaria x ananassa*), Beach Strawberry (*Fragaria chiloensis*), Musk Strawberry (*Fragaria moschata*), Alpine Strawberry (*Fragaria vesca alpine*), Wild Strawberry (*Fragaria virginiana*)

Growing: Strawberries generally need full sun and well-drained fertile soil in a temperate climate. They produce some of the tastiest fruit in the world. They are susceptible to pests and weather. Alpine does not spread like the other species do. Garden strawberries are the most popular, but they most be rotated every three years to stop diseases and fungus. Beaches and Musks work well as an effective groundcover.

Strawberry harvest grid:

Yield	Harvest Time	Tools	Harvest Readiness	Washing	Processing	Storage
1 pint per row foot	25 pints per hour	Bins	Commercial berries are picked before fully ripe	Do not wash	Free of blemishes and damage	Store in cooler overnight for delivery the next day

Sumac

Usage: Edible fruit

Species: Staghorn Sumac, (*Rhus* typhina), Fragrant Sumac (*Rhus aromatica*), Smooth Sumac (*Rhus glabra*)

Growing: Sumac enjoys partial shade to full sun with well-drained soil. It grows easily in any kind of soil. The berries are edible and are usually dried and then ground up to form a powder that is used to season rice or meat, and in some places it is used to make a drink like lemonade. The most effective way to contain sumac is with goats, as it tends to spread and will just grow more if mowed. Also, use caution as poison ivy, poison oak, and poison sumac are all part of the same family and have some similarities.

Sunflower

Usage: Edible

Species: Sunflower (*Helianthus annuus*)

Growing: Sunflowers need full sun and fertile, well-drained soil with regular watering. The seeds are harvested at the end of summer from the dried flower heads, and can be eaten raw or roasted, ground into butter, or the oil can be extracted for cooking. The leaves can be fed to cattle. Sunflowers are very easy to grow as long as they get enough sun.

Sweetfern

Usage: Ground cover, nitrogen-fixing legume, dryland tolerant

Species: Sweetfern (*Comptonia peregrina*)

Growing: The Sweetfern is not a fern, but rather a shrub. It enjoys well-drained soil and is hardy to cold. It enjoys partial shade to full sun. The leaves are used as an herb in cooking or for medicinal purposes. It also produces an edible fruit that may be eaten raw or cooked.

Sweet Woodruff

Usage: Wetland tolerant, edible

Species: Sweet Woodruff (*Galium odoratum*)

Growing: Sweet Woodruff is an herb that enjoys rich, moist soil and regular watering. It makes an excellent garden border as deer won't eat it. It has a strong scent that is used in potpourri and repels moths. It is edible but not in high doses. It is commonly used as an herbal medicine.

Taro

Usage: Edible roots and shoots, wetland tolerant

Species: Taro (*Colocasia esculenta*)

Growing: Taro can be grown in a paddy, like rice, to keep control of the weeds, but the water must be cool and flowing. However, a marshy place also works. It also enjoys a long, hot summer. To harvest, wait until the leaves turn yellow and dig them up. Make sure that you are eating the edible varieties and not the ornamental ones. The plant must be well cooked, as the raw parts can cause a burning sensation in your mouth and throat. To prepare them, soak overnight in cold water and then boil, bake, or roast like potatoes. In the United States, the Taro is also known as Dasheen and is cooked or dried and ground into flour. They can also be stored in live storage like potatoes. The leaves are also edible and often cooked like kale or used in traditional Hawaiian dishes.

Thyme

Usage: Ground cover, edible, bee forage
Species: Woolly Thyme (*Thymus psuedolanuginosus*), Citrus Thyme (*Thymus citriodorus*), Wild Thyme (*Thymus serpyllum*), Common Thyme (*Thymus vulgaris*)
Growing: Thyme is a perennial herb that enjoys full sun and well-drained soil. It tolerates drought and cold northern temperatures. It is used in cooking and herbal medicine and may be used raw or cooked. Woolly Thyme makes the densest ground cover. The leaves may be harvested throughout the season, as they will grow back quickly.

Tomato

Usage: Edible fruit
Species: Tomato (*Solanum lycopersicum*)
Growing: Determinate tomatoes, also called "bush" tomatoes, are varieties that only grow to about four feet high. They stop growing when fruit sets on the top bud. They also ripen all at the same time (usually over a two-week period), and then die. They may require a limited amount of caging and/or staking for support. Tomatoes should *not* be pruned or "suckered," as it severely reduces the crop. They will perform relatively well in a container (minimum size of five to six gallons).

Tomato harvest grid:

Yield	Harvest Time	Tools	Harvest Readiness	Washing	Processing	Storage
3–8 lb. per row foot	60–200 lb. per hour	Bins	Varieties vary on firmness, but usually when firm, heirlooms are softer	Wipe off and remove green tops	Free of blemishes and damage	Store in cool area with low humidity

Indeterminate varieties of tomatoes are also called "vining" tomatoes. They will keep producing fruit until killed by frost and can reach heights of up to ten feet, although six feet is considered the norm. They will bloom, set new fruit, and ripen fruit all at the same time throughout the growing season. They require substantial caging and/or staking for support and pruning and the removal of suckers is practiced by many but is not mandatory. The need for it and advisability of doing it varies from region to region. Experiment and see which works best for you. Indeterminate varieties are not usually recommended as container plants because of their need for substantial support and the size of the plants. Most urban farmers grow these tomatoes in a greenhouse by tying the plants to the ceiling with a string.

Turnip

Usage: Edible roots and shoots

Species: Turnip (*Brassica rapa*)

Growing: The entire turnip plant is edible, with the root eaten like potatoes, and the greens cooked like kale. They prefer well-drained soil, but they are extremely versatile and are grown in just about every climate and condition. Turnips also serve as animal fodder. They are harvested at the end of the season or left in the ground over the winter, or put in live storage.

Turnip harvest grid:

Yield	Harvest Time	Tools	Harvest Readiness	Washing	Processing	Storage
½ bunch per row foot	30–50 bunches per hour	Boxes	2–3 inches wide with healthy tops. White with no cracks or damage	Soak all bunches, wash 2–4 bunches at a time with sprayer	Free of blemishes and damage	Store in cooler

Vetch

Usage: Bare soil erosion control, nitrogen-fixing legume, ground cover

Species: Hairy Vetch (*Vicia villosa*), Bitter Vetch (*Lathyrus linifolius montanus*), American Vetch (*Vicia americana*), Wood Vetch (*Vicia caroliniana*), Tufted Vetch (*Vicia cracca*), Sweet Vetch (*Hedysarum boreale*), Milk Vetch (*Atragalus glycyphyllos*)

Growing: As vetch is often grown in conjunction with clover and alfalfa as a ground cover, it sometimes becomes part of a pasture that animals forage in. However, vetch is toxic over periods of time. If your animals eat too much of it for too long, they can end up with a nervous system disorder. Despite this, it is still useful because it is the most cold-hardy cover crop.

Violets

Usage: Ground cover, edible leaves

Species: Canada Violet (*Viola Canadensis*), Labrador Violet (*Viola labradorica*), Sweet Violet (*Viola odorata*)

Growing: These cute flowers have edible leaves, but not all are created equal. They grow just about anywhere, but local native species will vary in flavor. They act as a hardy evergreen ground cover. Sweet Violets are the sweetest and do the best where it is not too hot.

Walnut

Usage: Edible nut, wood and timber
Species: Heartnut or Japanese walnut (*Juglans ailantifolia* var. *cordifolia*), Butternut or White walnut (*Juglans cinerea*), Black walnut (*Juglans nigra*)
Growing: Walnuts enjoy full sun and grow 60–100 feet (18–30 m) tall and 50–100 feet wide. Black walnut improves the soil by building mineral content, unlike the other varieties, but they all taste delicious. The Black walnut grows from the Midwest to the East extending into the southern US, the Butternut grows in the Northeast radiating out from the Great Lakes, and the Heartnut grows in the coldest northern areas. Black walnut is considered the most valuable because it is the scarcest. Walnuts need fertile, moist soil. To harvest, wait until the walnuts begin turning from green to brown and start falling off the tree. The ones that are ready are partly brown and partly green. If they are completely black, they may not be usable. These must then be hulled by crushing them. Some people do this with their cars, by stomping on them, or using converted cement mixers. Once hulled, they must be dried for two months in a place with good air circulation. Place them in burlap or loosely in a similar breathable bag. After they are dry, they can then be cracked like any regular nut.

Water Chestnut

Usage: Edible, water plant
Species: Water Chestnut (*Eleocharis dulcis*)
Growing: The Water Chestnut is not a nut, but rather a water plant that grows an edible vegetable that is small and rounded, looking somewhat like a nut. These can be eaten raw, boiled, cooked, or pickled. They enjoy a long growing season and full sun. This plant should be placed around the shallow edges of a pond, with the water level staying at least two inches above the soil. At the end of summer, the leaves will die and then you simply dig up the plant. Rinse them off and use them accordingly.

Watercress

Usage: Stream bank erosion control, wetland tolerant, water plants
Species: Watercress (*Nasturtium nasturtium officinale*)
Growing: Although the Latin name is *Nasturtium*, they aren't closely related to Nasturtium flowers. This highly nutritious plant can be grown in wetlands or ponds, as long as the soil is fertile and water is abundant. It enjoys partial shade and warm temperatures. Harvest the leaves throughout the season and use fresh.

Wild Ginger

Usage: Wetland tolerant, edible roots and shoots, ground cover
Species: Wild Ginger (*Asarum canadense*), Shuttleworth's Wild Ginger (*Asarum shuttleworthii*)

Growing: A lovely groundcover that grows natively in most areas but still doesn't quite have the same flavor as cultivated varieties. Only the root is edible; the leaves are toxic. It is actually not related to gingerroot but is called ginger because it tastes and smells like ginger. They prefer moist, fertile soil, and shade. It is hardy in cold climates.

Wild Rice

Usage: Water plants
Species: Wild Rice (*Zizania aquatica*), Northern Wild Rice (*Zizania palustris*)
Growing: Rice growing has been described elsewhere in this book as a grain crop, but it can also be grown as an incidental water plant. Wild rice is actually not closely related to the common Asian rice (*Oryza sativa*). It is very nutritious. The seeds are simply thrown into a pond and sink to the bottom to sprout, and they enjoy full sun and cooler climates. They don't compete well with cattails and should be kept separate. The rice is delectable to birds and other foragers, so it may be prudent to protect it. Commercial growers drain their paddies to harvest the rice, but that's not an option for the small pond owner. Native people would use a canoe, but if you can't do that, just make sure the rice is planted within arm's reach of the shore. Then in the fall when the seed heads have fully developed and fall off easily, use a stick to tap them into a basket. These must then be laid out to dry in the sun for at least a day, and then parched. This is traditionally done in a cast iron pot over a wood fire, and must be stirred continuously until it changes from green to a darker color. This takes a couple of hours. Use just as you would other types of rice, but add three cups of water for every cup of rice.

Willow

Usage: Stream bank erosion control, wetland tolerant, water plants, bee forage, wood and timber
Species: Willow (*Salix L.*)
Growing: Willows are beautiful trees with many uses. The bark contains salicylic acid and is used along with the leaves in herbal medicine. It has tough, flexible wood used in making a variety of items. The plant grows easily from cuttings or broken branches that fall on the ground, and they grow in most climates. They prefer moist soil and are often planted on stream banks to stop erosion with their many tangled roots. In an urban area, these roots can cause trouble by destroying pipes and wires, so use caution. There are hundreds of species, so use the variety that is native to your area.

Yarrow

Usage: Bee forage, ground cover, edible
Species: Yarrow (*Achillea millefolium*)
Growing: Enjoys partial shade to full sun and provides beautiful ground cover that gives a home to beneficial insects. It has many herbal medicinal uses and benefits the soil. However, too much yarrow intake can cause some severe digestive problems, so take in moderation.

5 | Animals

▲ The urban nanny goat.

BEFORE YOU GET ANIMALS

"All goats are mischievous thieves, gate-crashers, and trespassers. Also, they possess individual character, intelligence, and capacity for affection which can only be matched by the dog. Having once become acquainted with them I would as soon farm without a dog as without a goat."
—*David Mackenzie,* Farmer in the Western Isles *(1954)*

Animals are much more complex than plants and have a lot of emotional and physical needs. That being said, farm animals also don't have a huge learning curve. If you are a caring person who can recognize a living being's needs, you can raise an animal. These are the basics that you absolutely need to know before getting started. In addition, it's a really good idea to find someone who has the kind of animal you want and just spend a couple of hours picking their brain. In this case, you really can only learn by doing. Just start small and add more animals slowly so you don't end up in trouble. It's important to note that there is no way this book can cover everything you need to know about animals and how they fit within an urban farming system. This chapter is just a guide to help you make better farming decisions.

Production Planning

Animal	Number	Output	Harvest
Chicken	6 hens	180 eggs/month	Spring & summer
Meat ducks	2 females/1 male	25 ducks/year	
Egg ducks	6 females	180 eggs/month	Spring & summer
Small goat breed	2 females	15 gal/month	Spring & summer
Honeybees	1 hive	5 gal/year	Late spring or summer
Rabbits	3 does/1 buck	75 per year	Spring & fall

Animal Behavior

Understanding how an animal thinks is important for your well-being and the animal's, even if the animal is a wild creature you encounter in the wilderness. Humans have the tendency to think of everything as a human—or "anthropomorphize" things. The only creatures that we can really do this with are house pets like cats and dogs. Animals learn by cause and effect, and it takes time and repetition to learn behavior based on results. Much of what an animal does is based on instinct. I believe that an animal does feel attachment to a human, but it is also important to remember that the love the dog feels is doggy love. It is an instinctual love that is descended from the wolf pack, and that is the way the dog is relating to you. No matter how long an animal lives with you, it will still react in the ways that it has evolved to react. Anything that an animal does in relation to you will be your fault.

Biosecurity

While most small farmers don't have to worry too much about biosecurity, you still need to protect yourself against foreign health risks to your animals. It is just common sense to protect their health by preventing exposure to diseases from other animals, especially lately with avian flu, mad cow, hoof and mouth, etc. Small farms and homesteads tend to have hardier livestock. You can follow these precautions to help raise your animals to be hardy and build their immune system just as you would your children.

1. *Limit access.* Unless you are making money from agri-tourism, you should limit the number of people who come in and have access to your animals. If you must have lots of visitors, they should clean their shoes before walking around. It is especially dangerous to have guests who have been overseas less than a week before; they should wait longer before being exposed to your animals.

2. *Change your clothes.* When you go off the farm, and especially if you have visited an animal show, auction, or another farm with animals, change your clothes and shoes before handling your animals. Pick one pair of shoes that you only wear on your own farm, and another pair for going places.

3. *Purchase carefully.* When you are going to buy a new animal, be very careful whom you are getting it from. Do your research and buy only from breeders with a good reputation. Make sure you know how healthy their herd is. Sometimes it's better to buy from someone who has had animal health problems but fixed them, rather than someone who has no idea how healthy their herd is.

4. *Quarantine.* When you buy an animal, or if one of your animals leaves the farm for any reason, quarantine it for at least twenty-one days. When it is by itself, observe it for disease so there is no risk of it spreading anything to the rest of your livestock.

5. *Keep clean.* You don't want mice, rats, and other pests to eat your animal's feed. Not only do they make a big mess, they also poop in the food and spread disease. Use prevention to keep these critters out by keeping food in rodent-proof containers and keeping the area clean. You should also clean anything that comes in contact with manure, and it is recommended that you disinfect it.

6. *Observe.* Keep a close eye on all your animals every day so that you will notice any unusual behavior or symptoms, such as blisters around the nose and lips, blisters around the hooves, or staggering. That way you can catch a problem right away and hopefully stop it from spreading. If any of your animals dies for an unknown reason, take it to a vet and have him find the cause.

First Aid Kit

Note: Everything in the kit should be kept cleaned and sterilized.

First aid kit checklist:

- Toolbox
- Restraint equipment (for individual animals)
- Digital rectal thermometer with string and alligator clip
- Stainless steel bucket
- 16, 18, & 20 gauge hypodermic needles, 1–1.5" long
- 5cc, 12cc, & 60cc syringes
- IV complex hose
- Obstetric chains and handles
- Neonatal resuscitator
- Neonatal esophageal feeder
- Feeding tube (sized for individual animals)
- Small refrigerator for medicine
- Flashlight
- Funnel and soft rubber tubing
- Cotton balls
- Q-tips
- Cotton rags

Wound kit checklist:

- Rubber gloves
- Antiseptic cream
- Betadine cleansing solution
- Hydrogen peroxide
- Nonstick gauze pads
- 4" x 4" standard gauze pads
- Cotton roll
- Vet wrap
- Adhesive tape

RABBITS

How They Fit into the System

In a permaculture system, bunnies are not kept as pets. They provide meat, manure, and fur. Unlike other animals, they can never be allowed to roam the gardens because they will eat everything and multiply beyond control. Instead, they live in a hutch with a mesh floor so that their droppings will fall down for easy cleanup. A worm bin can be placed underneath for

▲ Rabbits aren't fussy about their living space.

even easier cleanup. Rabbits eat grass, leaves, twigs, hay, vegetables, and kitchen scraps, and they are especially practical for urban farmers because rabbit hutches are legal in most places. Any rabbit breed can be eaten, but there are rabbits that are designed specifically for meat and grow to be much larger, such as California, New Zealand, Champagne d'Argent, or Florida White rabbits.

The nutritional value of rabbit is not as high as chicken or other meat staples, but since they are very self-reliant and quiet they make a valuable meat source. They need a constant supply of fresh water and the wire mesh of their cage should not be larger than half an inch. The hutch should be kept clean and dry and be sheltered from the weather.

Breeding

Rabbits breed very prolifically and easily, but may not always be the best mothers. Small breeds should not have babies until they are six months old, and large breeds not until they are nine months old. Males and females should be kept separate and only put together for breeding for a short time under supervision by putting the doe into the male's cage so that he notices her. It's a good idea to have him try again about eight hours later just to make sure. Once she is pregnant, she should be provided with a nesting box and soft material such as hay or down. After the babies are weaned, the males and females should be separated so that they will not breed too early. They become very territorial and will need their own cages shortly anyway. It should be noted here that while siblings should not mate, it is common practice among rabbit breeders to allow mother/son and cousins to breed. Don't breed a rabbit with defects or illness, and choose a good breeding pair promoting the best traits. Around thirty days later, the doe will *kindle* or give birth.

▲ Newborn rabbits in a clean nest of hay and fur.

A doe needs a nesting box, which is a low-sided wood box with a tall back and a roof awning that comes down over the back, in her cage at least a few days before she kindles. She will pull her own fur out to make a nest. If she doesn't, you can gently pull small bits of fur out all over her body. A couple of days before birth her fur will loosen and this will be easy to do. You can also add clean straw for even more warmth. Sometimes does don't give birth in the box, and you will have to move the babies back into the nest. They will probably be too cold to survive, but you can try to warm them up. If you are able to save them, put them back in the nest in a little hollow of fur and cover them. Sometimes does eat their young, which can be caused by anything from stress to poor nutrition to just disturbing her when she

was about to give birth. She may just be a terrible mother. Sometimes does don't feed their young, or they step on them and kill the babies accidently. It is possible to try to feed them but chances are they won't survive. Since she can have another litter soon, it may not be worth the trouble of feeding them every two hours. It is typical to breed a doe every six weeks and wean the babies around five weeks old, and with such an intense breeding schedule you may be able to raise three hundred pounds of meat.

CHICKENS

How They Fit into the System

Chickens are foragers, so the best setup is to have them in a foraging area,

▲ A backyard chicken coop that provides great predator protection.

but still close enough to your house to be convenient. They provide meat, eggs, feathers, fertilizer, pest control, and weed control. The chickens live in a coop that has an attached pen that runs along the border of the gardens and, if possible, an orchard. It should also have a second pen for chicks. The trees and ground in their pen should be well mulched with straw, corn stalks, sawdust, yard waste, or bark, with wire mesh around the trees holding the mulch in. The chicken run can be planted with fruit trees (which will drop fruit on the ground), grains, corn, sunflowers, and greens. When weeding the garden, the waste can just be thrown over the fence into the chicken run. The run is divided into several pens planted in succession so the chickens can be rotated when the plants are ready, and each pen also has a log on the ground. The log is left to sit for a while, then flipped over to reveal all the pill bugs and worms. The fence dividing these pens and keeping the chickens from your other gardens should be at least five feet high.

Chickens can't be let into the mulched gardens or orchards that are used for food production because they'll destroy the mulch. However, you can let them into an unmulched orchard. When the orchard is young, only let bantams or very small chickens forage, where they will eat dropped fruit and weeds and fertilize the soil. There should be no more than fifty to one hundred chickens per acre in the orchard, or in a pasture with other breeds. Twenty-five light breed hens producing eggs will eat a quarter pound of food per day, and even more in a heavy breed. If you are using chickens as a cash crop, you may want to grow three hundred chickens per acre on a well-planned forage pasture. No other animals should be with them.

Alternatively, and this is especially the case for urban dwellers, chickens can live in a *chicken tractor* or *ark*. This is simply a small house for fewer chickens that is built to be moved around a yard or pasture. There are a myriad of designs and even commercially available chicken tractors made out of plastic that can be moved around a pasture in order to let grass grow again. See the Chicken Tractor section for how to build one.

The pasture for a tractor or pens can and should have shrubs such as pigeon pea, berries, and fruits, or plants like acacias, clovers, grasses, chicory, comfrey, and dandelions. Insects and larvae can be introduced by having a large manure pile or mulches, or you can grow your own crickets. You can give them cottage cheese or other dairy for more protein. They also need ground shells for minerals and grit to help them digest their food, which can come from their own eggshells, or from mussels or snails that you raise. If you don't provide this, they might eat their own eggs. Gravel, sand, and pebbles can also be put in a container on the ground for grit. See the section on ducks for a homemade poultry feed formula.

Eggs

The general consensus among free-range chicken owners is that you need at least one hundred chickens to make any kind of money from the eggs. A hundred chickens that lay a decent number of eggs, up to three hundred each per year, will give you thirty thousand eggs. These are usually sold for a couple of dollars (US) per dozen in a rural area and slightly more in a suburban location. If you take away your own share, you could make a couple of thousand dollars per year on them. This kind of profit is only possible if they are free range; otherwise, the money will be eaten up in feed costs.

Check your local laws regarding selling eggs. Usually, law requires that the eggs be clean and if you are selling more a certain number, they must be *graded,* or given a letter grade based on their quality. They don't usually have to be washed, or put in a new container, but they do have to be refrigerated. Selling eggs in a free-range setting isn't an endeavor that will pay your mortgage, but it will provide some extra income, especially since your chickens will be producing more eggs than you can use.

Chickens *molt,* or lose their feathers, once a year and usually in the fall, and they also stop laying eggs. The feathers will start falling from the neck, then the breast, thighs, and back, then the wings and tail. Molting is usually triggered by having less sunlight from shorter days. By using a light that is connected to a timer in the coop you can keep egg production up. If your chickens molt in the summer, it may be caused by stress, such as a food or water shortage, disease, cold temperature, or sudden lighting changes. Chickens also stop laying just because they get old. Their comb, vents, and wattle become shrunken and pale, their body will be smaller, and as time goes on, their vent, eye ring, and beak will become yellow.

Chicken Coop

The coop must be at least 3 square foot (0.3 m) per chicken. A manageable size is 7 x 7 feet (2.1 x 2.1 m), or 49 square feet (4.5 square m), which is enough space for sixteen chickens. A small chick-house can be attached to the side that is fortified from predators. Roosts are long poles or boards

Chicken Plants

These can be planted in your pens and pasture for your chickens to eat from. See the chapter on plants for more information.

Alfalfa	Fava beans
Amaranth	Fennel
Autumn olive	Fruit and nut trees
Barley	Honey locust
Buckwheat	Oak
Chickweed	Oats
Chicory	Quinoa
Clover	Russian olive
Comfrey	Rye
Corn	Siberian pea shrub
Cucumber	Stinging nettle
Currant	Sunflower
Dandelion	Swiss chard
Elderberry	Vetch
	Wheat

at least 18 inches (45 cm) from the wall and low enough for them to fly up to. Nesting boxes are square boxes around the wall about 18 inches (45 cm) from the ground, at least 12 x 12 inches (30 x 30 cm) in size. These boxes can be attached to the outside of the coop with a lid so that you don't need to go in to disturb the chickens every day to get the eggs. The general rule is one box for every two hens. The chicken door should be 12 inches high (30 cm). In a cold climate, the coop needs extra insulation and should be able to be closed up at night. The following list of guidelines will help your chickens stay healthy:

Litter: Spread a moisture-absorbing cover such as wood shavings at the bottom of the coop. It should be at least four inches deep, loose, and dry. The coop should have proper ventilation and few water spills. Instead of cleaning it out once a week, you can pile it up until it is two feet deep.

Cleanliness: All houses and equipment should be disinfected before any new chickens arrive. Remove wet litter, moldy or wet feed, dirty water, or clean out nests when droppings get in. Once a year the house should also be cleaned and painted.

Water: Chickens should have fresh water every day, and it should always be available to them. The best way to do this is buy a chicken waterer that is designed to hang from the ceiling—it should be kept off the ground.

How to Keep Chickens Healthy:
- Disinfect the coop and equipment before you bring in new chickens.
- Keep bedding clean and dry, cleaning it out once a week.
- Clean and disinfect the coop once a year.
- Provide fresh water every day, making sure it is always available.
- Separate old birds from young birds to prevent the spread of diseases.
- Provide a small amount of crushed shells every day.
- Supply grit and keep water and food containers clean to prevent worms and parasites.
- Prevent cannibalism by providing adequate space and nutrition.
- Ensure that birds get needed vitamin D. In a northern winter, you may need to provide cod liver oil.

Young birds: Keep the young birds away from the old birds because they can catch diseases they aren't immune to. They may also get pecked to death.

Windows: Put the window on the south side in northern climates, on the side of the roof that is lower so that the roof overhang shelters the window. The window should be able to be opened to provide adequate ventilation.

Doors: You will need a human door and a chicken door (about one square foot) that can be shut from the outside. You can close the little door at night for protection but you'll have to get up early to open it.

The yard: Some people let their chickens roam the property, and others make a yard fenced with chicken wire. Smaller chickens need a fence of at least five feet. They enjoy litter such as straw, leaves, corn stalks, or cobs to scratch in.

Chicken Tractor

The most common small design for a chicken tractor is an A-frame, with a triangular house and a totally enclosed triangular pen covered in chicken wire. The other common design is a rectangular pen. The house takes up one-third of the total space and has a square cut out for the chickens to go out into their tiny run. A side wall or roof of the house is hinged so it can be opened up for collecting eggs and cleaning.

Another feature that makes moving the chickens around much easier is wheels. The wheels must be set very low to the ground, as any gaps around the bottom of the pen can allow predators to get in. The solution to this is an extra wood barrier that is set around the pen that is simply removed when you have to move the chickens.

The Ultimate Guide to Urban Farming

▲ **The A-frame chicken tractor.**

Breeds

There are so many different types of chickens that when people start to shop for their first batch they are often overwhelmed. In North America, most of our eggs come from white hens that lay white eggs, but there are hundreds of other breeds to choose from. These are divided into two categories: *light* or *heavy*. The light breeds can fly short distances. They aren't good mothers and aren't as hardy as the heavy breeds, but they are excellent at foraging and don't usually need supplementary food. The heavy breeds don't fly, are better mothers, and usually lay brown eggs. They tend to be hardier and lay eggs longer during the season, but they aren't as good at foraging. Some breeds are also much nicer than others, cutting down on pecking order problems. When chickens see a weakness in a bird either because of size or injury, they often gang up on it and peck it repeatedly, sometimes to death. Orpington is one of the most

popular dual-purpose breeds. They are fairly laid-back and easy to handle, making them a good choice for a beginner.

Feeding Chickens

Twenty-five light breed hens producing eggs will eat five to seven pounds of feed per day. You can buy a commercial feed (preferably with 15–16 percent protein), or you can mix your own by combining ground corn with proteins and minerals. Chickens can also eat kitchen waste, but only once a day, and only enough to eat in 5–10 minutes. Onions and fruit peels can make the eggs taste funny. Chickens also need ground eggshells or oyster shells and grit to help them digest their food; it should be sprinkled on top. Remember to change the feed gradually over a week, increasing new feed from one-quarter to one-half to three-quarters. What most people do is feed lots of milk, as much grain as they want, a whole pile of scraps, then sprinkle the egg or oyster shells on top, and provide a continuous supply of forage and lots of water. It doesn't have to be scientific. Gravel, sand, and pebbles can be put in a container for grit.

Types of feed:

Commercial feed: Usually contains chicken parts ground up and labeled as "protein." For chicks, crumbles are best. For older chickens, pellets or mash work well. They come in packages formulated by age so you just pick the package suitable for you.

Starter feed antibiotic: A commercial chick feed that contains antibiotics. It should only be used for a week because the birds can become dependent on it, and it is expensive. Any feed with antibiotics should not be used for birds that will be butchered in the next week.

Homegrown chick feed: Mix two parts finely ground wheat, a little corn, and oats; one part protein: such as fish meal, meat meal, canned cat food, hardboiled eggs, yogurt, cottage cheese, worms, bugs, or grubs; and one part greens: alfalfa meal, alfalfa leaves, or fresh greens, such as finely chopped lettuce. You can add wheat germ, sunflower seeds, linseed meal, etc. Their diet should be 20 percent protein. Besides sand for grit, sprinkle ground oyster shells or egg shells.

Homegrown adult feed: The best meal is about the same as for chicks, but you can use slightly less protein, 15–16 percent. Chickens can eat peels, sour milk, pickles, meat scraps, rancid lard, overripe and damaged fruits and vegetables, pods and vines, table scraps and stuff to throw out of the fridge. They won't eat onions, peppers, cabbage, or citrus fruit. They shouldn't eat moldy food.

Health Care

Vaccinations: Birds should be vaccinated before twenty weeks for infectious bronchitis and Newcastle disease. If you buy pullets, they should also have been vaccinated for Marek's disease in addition to the others. Consult your local veterinarian for other vaccinations needed for your area (some places have fowl pox or other diseases).

Parasites: Mites and lice are the common external parasites, and roundworms, cecal worms and capillary worms are the most common internal parasites. Cleanliness and management will prevent them. If you get external parasites, dust with wood ashes, powdered sulfur or diatomaceous earth, or dip the birds in 2 ounces of sulfur and 1 ounce of soap per gallon of water. The affected birds must be quarantined in order to prevent spreading. Worms can be helped by feeding them garlic and lots of grit.

Cannibalism: Birds naturally eat each other, but it is caused by stress, overcrowding, not enough food or water, space, malnutrition, the wrong temperature or the sight of blood on a chicken. Make sure you are not causing the problem, and then debeak the murderous bird. Use a sharp knife or toenail clippers to cut off the tip of the beak (but not far enough to cause bleeding).

Injuries: A broken leg can be splinted with a popsicle stick and masking tape; however, if the chicken does not have a disease, you may want to go ahead and use it for meat.

Egg-bound chicken: This is when an egg gets jammed in the chicken. She will strain to lay it but won't be able to and will look constipated. Pour warm olive oil in her vent (her rear), and then try to rotate the egg out yourself.

Prolapsed vent: This can happen when a pullet lays too early. The end of the oviduct inside the bird will hang out of the vent. Wash the protruding tissue with warm water and a mild antiseptic, then lubricate with petroleum jelly. Push the mass of tissue back into the vent gently, dry her off, and then separate her from the rest. Feed her lots of greens and fresh water (no grain) to slow the egg production. In seven days she will be OK, but if this happens often you may want to use her for meat.

Impacted crop: Also called crop-bound, this is when the chicken eats something wrong and it gets stuck in the throat so it can't eat. It will have a fat, soft throat and move its neck convulsively. Pour a teaspoon of olive oil down the throat and gently massage the neck, working the contents up and out of the mouth. Give the bird only water for twenty-four hours, then feed solids. If this happens often the neck muscles were injured, so you will have to turn it into meat.

Vitamin D: A bird that doesn't get enough vitamin D won't thrive, will lay thin-shelled eggs, have leg deformities or other problems. In winter, if you live too far north to let them run in the sun, you can give them cod liver oil.

Only really big commercial farms have lots of diseases. Usually a small flock has very few. You can't eat a sick bird, so prevention is the key. If you do find a sick chicken, isolate it, keep it warm and feed it well. Some things can be vaccinated against, but usually a small flock doesn't need it unless your local area is having an outbreak and a vet recommends it. If the bird dies, you will have to bury it.

Diseases that can be vaccinated:

Marek's disease: Symptoms of this disease include leg paralysis, drooping wings, and weight loss. Birds may have tumors on their internal organs. A bird may carry the disease but not show symptoms, but other birds may die from it. The vaccination does not work after exposure to the disease for three or more days, and infected flocks will be contaminated forever.

Infections Lyrngotracheitis: Birds gasp for air and cough up blood, and it is frequently fatal. To avoid this, vaccinate after four weeks of age and administer a booster every year.

Fowl pox: People get chicken pox, chickens get fowl pox. Humans can't get fowl pox. Birds get round scabs on unfeathered skin, fever, and weight loss. Birds that get it in the mouth sometimes die of starvation or suffocation. It is spread by insect bites or through wounds. All birds in the flock should be vaccinated in early spring or fall, with a yearly booster.

Respiratory diseases: Viruses such as Newcastle's disease, infectious bronchitis, mycoplasmosis, turkey and chicken coryza, and avian influenza all have similar symptoms including eye swelling, runny nose, coughing, and poor weight gain. Get a blood test, bacterial culture, and virus isolation to find out what your birds have.

Gathering Eggs

1. Gather eggs at least once a day (three times is recommended) and clean out the nests once a week. If you leave them too long the chance of breakage

is higher, which can cause the bad habit of egg eating. Separate dirty eggs from clean eggs.

2. To clean dirty eggs, use the hottest water you can tolerate to prevent any microbes from entering the pores of the shell. Do not soak the eggs, and if you don't have hot water, don't wash them.

3. Use nonfoaming and unscented detergent (if it has a scent the egg will absorb it), such as dishwasher or laundry detergent or Borax, and wear gloves. Rinse off with clean water.

Molting

As mentioned previously, molting is a natural occurence that can be caused by stress or less sunlight when your hens won't produce eggs. It is possible to force a molt, which can make the hen's production life last longer, if you do it at about fourteen months old. This simply gives them a much-needed period of rest.

Forcing a molt:

Day 1: Turn off the artificial lighting, so that the chickens are only getting about eight hours of natural light a day. Keep giving them water, but remove all feed for ten days.

Day 11: Give the chickens a full feed of cracked grain for two to three weeks.

Two to three weeks later. Feed the normal laying ration and turn the lights on again. The chickens will be in production in six to eight weeks.

"Retiring" a Hen

You can tell a chicken has stopped laying if the comb, vents, and wattle are shrunken and pale. Their body will be smaller and their pubic bones will be close together and possibly covered in fat. Yellow

coloring will gradually return to the vent first, then the eye ring, earlobe, beak, and shank. Unfortunately, her laying life will be over and she is no longer pulling her weight. It's time to butcher her and replace her, but unfortunately, she won't taste very good and she can only really be boiled into soup. Most serious chicken owners retire their hens by two years, at that point they won't be as tough as a three-year-old hen but still not laying as they could be.

Incubating Eggs

You can purchase eggs from any agricultural store or dealer. Almost no municipality allows you to own a rooster because of the noise they cause. Roosters are very loud and they crow at all hours of the night and day, so it will be impossible for you to fertilize your own eggs. Considering how much trouble a rooster is and how unreliable they are at fertilizing, the eggs from the local dealer are really the ideal option anyway. Rather than incubating eggs you also have the option of purchasing chicks, which is described in the following section.

Electric incubators are cheap and easy to use, but you can make your own. Use a Styrofoam cooler or a cardboard or wooden box with a glass or plastic top. A wood incubator should be 11x16 inches, 11 inches high with a hinged front door. Drill ⅜-inch holes on each side, two near the top on the 11-inch sides, and two near the bottom on the 16-inch sides, for circulation. Make a tray of wire mesh on a frame 2 inches from the floor of the box, and put a water pan under it. Place the eggs on the tray and put a thermometer in with them. Use a 40-watt bulb to keep the eggs at 99.5°F at all times. Keep the pan of water full so that the air remains humid. It

is very important to turn the eggs gently, a quarter way around, three or five times a day (never an even number of times, or the chick will lie on the same side every night). Mark an *X* on each egg to keep track. After ten days, make sure the large end is higher than the small end. After eighteen days, stop turning the eggs, and raise the humidity in the incubator. Most eggs hatch between nineteen and twenty-two days. When a chick pecks a hole from the inside, it is called pipping. The chick may start to pip on the eighteenth day, but it won't actually do it until there is a hole showing. Whatever you do, *don't* open the incubator, don't even touch it. Don't help the chicks get out of the eggs; they must do it on their own or they might die.

Candling an Egg

Some eggs may not work, and then they can rot and even explode. To find out if you have an unfertilized egg, make the room dark and use a very bright light behind the egg to look through the shell. If you see a clear egg, it is not growing. If you see a dark haze or gray clouds, it is rotting. If you see a dark red circle and no veins, the embryo died. If you see a small dark center and a network of veins, the egg is good.

Raising Chicks in a Brooder

A brooder is simply a box with a light that keeps chicks warm. You can buy a brooder, or you can make one by using a box with a red heat lamp and a thermometer. The bulb should be low-wattage, and the temperature should be about 95°F (35°C). A box 30 inches square (76 cm) with a 69-watt bulb can brood fifty chicks. Put your hand down to test the heat. If it is uncomfortably hot, the lamp is either too close or too hot. You may need to switch to a bulb with a lower wattage. Check the chicks two or three times per night the first week. If they are cold, they will huddle under the light, and if they are too hot they will stay near the walls. If they are content, they will look and sound

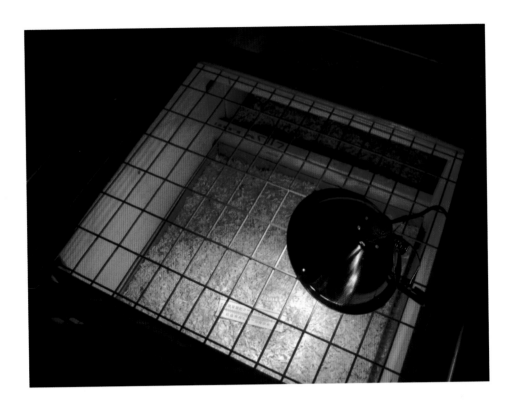

like it by cheeping happily. Decrease the heat 5°F per week, so that by six weeks it is 70°F (21°C). Then you can turn the heat off unless it gets chilly. As the chicks get older, tape another box next to the first and cut a door. Hang heavy cloth in the door and the other room will be cool, so the chicks can run in and out. The light should be red or green and very dim, as a sudden adjustment to total darkness could kill them. Chicks also need water and bedding. You will need one gallon (4 L) of water per fifty chicks. Keep the water clean and full at all times, and make sure it is at room temperature, not cold, although you can't put the water right under the heat lamp or it will be too hot and just evaporate. The first week use burlap or cloth rags (with no loose threads) laid out flat for bedding, over a layer of newspaper. Then, when they have learned what food is, graduate to a thick layer of black-and-white shredded

newspaper, hay (not straw), or wood shavings. If you do use wood shavings, the pieces should be too big to fit in a chick's mouth. Stir the litter every day, and remove wet spots to prevent spraddle legs (legs turning outward) and infection.

If you hatched your own, don't give them food at first. Wait until they start pecking at the floor; then give them food. If you bought them, have food ready because they will be three days old. For the first week, give food on a paper plate, egg carton, cardboard, or some other surface that will bring the food up closer to their eye level. Once they figure out what food is, put the food in a tuna can or a trough—the container should be difficult to walk in and scratch food out of, but short enough to reach in. Each chick will need one inch of feeding space until they are thirty days old, then they will need three inches. Don't put the feeder right under the heat lamp. Fill the

The Ultimate Guide to Urban Farming

feeder only halfway full to prevent them from throwing the food out. Clean out old food each time you fill it, and keep it full all the time. Chicks need small grit to help them digest food so sprinkle sand on top of feed.

Chick Dust and Moving Chicks

Chick dust is a powder that comes off the droppings when they dry that can get into your lungs and could potentially even cause lung disease. When birds start to molt, they can also release down into the air. Don't keep chicks in the house for very long. When they are four weeks they can stand on anything, so if you have a separate room in the coop (not partitioned with chicken wire, which they can walk through), transfer the heat lamp and box to it. When they are six weeks you can remove the box, and when they are ten to twelve weeks you can put them with the other chickens (if they are big enough and weather is warm). These small chickens are called pullets and should be moved before they start laying. If the weather is warm, after one week you can let the chicks out on grass to run, but they get tired easily and will need to come back to get warm.

Chick Problems

If your chicks get diarrhea, it is a sign of coccidiosis. The risk is higher if the birds are overcrowded or hungry. If they get it, add 1 tablespoon of plain vinegar to their drinking water. If a chick dies, remove it immediately so the coccidiosis doesn't spread. Some chicks get a *pasted vent*, which is the term that describes when a chick's back end becomes dangerously clogged with droppings. Use lukewarm water to remove the droppings, then rub petroleum jelly on the area, and make sure the chick is completely dry. Petroleum jelly can also be put on chicks that are getting pecked.

DUCKS

How They Fit into the System

Ducks are the gentlest and most versatile poultry. They eat algae and weeds from ponds, slugs, snails, grubs,

▲ Ducks can be quite happy with this minimal amount of water.

soft greens and grasses, water plants, small tree greens, and grains. At the same time, they fertilize the soil and the water, which improves fish production. They like to walk on small plants and they eat some of them, too. Therefore, they work better in a well-mulched area with plants that are well established. They also need less care and feeding than chickens, although they need more planning. While a few bantam chickens can be thrown in a greenhouse, a flock of ducks require lots of water and

---

Homemade poultry chick feed:

30 percent grain: finely ground wheat, a little corn, and oats

20 percent protein: fish meal, meat meal, yogurt, cottage cheese, worms, bugs, and grubs

50 percent greens: alfalfa meal, alfalfa leaves or fresh greens such as chopped lettuce

Extras: wheat germ, sunflower seeds, linseed meal

Sand for grit

Ground shells: mussel, snail, oyster, or eggshells

Homemade poultry adult feed:

30 percent grain: finely ground wheat, a little corn, and oats

15 percent protein: fish meal, meat meal, yogurt, cottage cheese, worms, bugs, and grubs

55 percent greens: alfalfa meal, alfalfa leaves or fresh greens such as chopped lettuce

Extras: wheat germ, sunflower seeds, linseed meal

Sand for grit

Ground shells: mussel, snail, oyster, or eggshells

---

grazing. If you have other animals that need to drink from watering troughs or ponds, the ducks must be separated from those water sources or they will make them too dirty. The general rule is that you can keep twenty-five ducks per acre of pond surface; however, ducks don't *need* a pond necessarily. For an urban farmer, this means that the ducks need a place to splash around, at least a few inches deep. If you have a small pool three feet across and twelve to eighteen inches deep that ducks can dip their heads and feet in, the water will prevent diseases and feed them with bugs. Most city duck owners set a bathtub in the ground that the ducks can play in. If the duck area is next to a garden, you can open it up now and then so they can eat the slugs and pests, but only when the plants are at least as big as the ducks.

Duck Care

If the ducks have adequate water, a grassy yard with new grass, and a forage garden with bugs in it, then you won't need much extra feed. Ducks need young grass to eat, and if their pasture is too small and unvaried, they will quickly destroy a grassy backyard. If you must give them additional food, wheat is the best grain for ducks, and it goes well with oats. Hard round fruits and vegetables need to be crushed for them first. Liquid milk and hard-boiled eggs are good sources of protein for laying birds, and all ducks need calcium from eggshells or seashells, and grit. Ducks, unlike other poultry, need a little more niacin in their diet, but lots of fresh greens or peas should be enough to provide them with what they need. In turn, they will give you eggs and meat, feathers, pest control, and fertilizer.

Ducks are social creatures and need a flock to be happy, so you'll need to have at least two ducks or more. Each duck needs four square feet of housing. During the day this can be a three-sided shelter near the pond, but at night they need to be kept away from predators. They can be put in a simple shed. This shed should have a door for people so that you can harvest all the valuable fertilized straw. Ducks can usually fly and need to be clipped, or they will fly around your whole property or even leave completely. Use big scissors to clip off the ends of long feathers of one wing when they first grow, and after each molting after that. Don't cut during the molting or you may cause fatal bleeding.

In the winter, they will run out of forage and can quickly turn the area around the pond into mud. Throw down another layer of mulch, such as fallen leaves or hay, as they stir up the earth. Ducks can tolerate freezing temperatures as long as they can still run back into their three-sided shelter away from the wind when they need to.

There are several breeds that are popular with duck owners that are better for eggs, meat, or both. Khaki Campbell is the most popular and was the first domestic duck breed, followed by Indian Runners. They produce just as many eggs as chickens, but are not good meat birds. Meat breeds include Muscovy, Rouen, and Pekin. To find a breed that is *dual purpose* or works well for meat and eggs, you will have to look to history and pick a heritage breed. Heritage breeds are types that were raised by small farmers over hundreds of years and are now not as commercially viable. Because of this, they are dying out. Ancona, Appleyard, Buff, Magpie, and Saxony are good dual-purpose breeds. Saxony is probably the best of these for their foraging and egg-laying ability.

Duck versus Chicken

Ducks may be nicer, but they are noisier. They do better around children than chickens, but make sure that you always act and talk gently around them or they will startle easily. Duck eggs taste different than chicken eggs, and ponds will add different flavors. You can cook duck eggs just like chicken eggs, except they can be a bit tougher—when frying, add a bit of water.

Breeding

Duck males and females are difficult to tell apart. The female will have a loud, raspy quack, and the male will be a bit quieter or sometimes silent. Some male ducks will also become very protective of the females. If you have a motherly duck, it is best to let her raise her own ducklings, as she'll do a better job than you. Ducks will start laying in the spring around six to seven months old and keep laying for three years or longer. They always lay in the morning and are very scheduled, so let them out of the barn after ten in the morning and then lure them back into the barn in the evening with a handful of grain. A mother duck and her ducklings should be kept separate from the other ducks until the chicks are six to eight weeks old. It has been recommended that very young ducks under five weeks old should also be kept away from the pond.

Brooding is about the same as for chicken, but the holders will have to be bigger to accommodate the bigger eggs. The temperature should be 99–100°F for forced air, or 101–102°F for still air. The eggs need to be kept very humid and moist. To do this either spray lukewarm water on them once a day or put a large sponge into the water pan and sponge the eggs gently with warm water when you turn them. Stop turning the eggs on the twenty-fifth day but keep spraying them with water until they pip. It will take twenty-eight days (except for Muscovy, which is thirty-five days), and on that day give one last spray and leave them alone.

Caring for Ducklings

Each duckling will need 1½ feet of floor space until they are seven weeks, then they will need 2½ square feet. For every thirty ducklings, you will need a 250-watt heat lamp, a little higher than you would need for chickens. Leave the light on all night, and make sure the outer edge of the heat circle is 90°F. Reduce the heat 5°F every week until it is at 70°F. The first two weeks the ducks should not be allowed to get wet, but they should have a drinking trough at least 2 inches deep, and only 1 inch wide. At four to six weeks, turn off the heat unless it is super cold. By four weeks, they should have started growing feathers so they can go outside in the daytime. Any other duck care is just like chickens.

Instead of chicken feeders for ducklings, use small box tops near the heat lamp, and then graduate to rough paper, such as the bottom of grocery store paper bags. You can use commercial chicken starter feed if it doesn't have antibiotics and it is recommended for ducks. To make your own feed, give them cooked oatmeal with water for breakfast, scrambled eggs with water for lunch, and whole wheat bread with water for dinner, with some cut up greens. Gradually give them more and more greens, and at two weeks start giving them some grit.

PIGEONS AND QUAIL

How They Fit into the System

Pigeons are kept in tall cages that you can walk around in. Quail can be kept in much smaller cages and up to six can be raised in a square foot (0.09 m), although for our purposes we would want to give them more space than that. Quail can also live in the greenhouse because they don't eat the plants; pigeons eat seeds and grain, and quail eat insects. They provide eggs and meat, and like rabbits can be legally grown in the city. For people who live in urban locations and are not able to raise chickens, pigeons and quail are sometimes allowed (although not necessarily in the quantities you will want to raise). Quail are considered wildlife, and in many places you may need to get a game bird license. A breeding pair of pigeons can produce twelve *squabs*, or baby pigeons, per year. Squabs are considered a gourmet dish and are incredibly easy to raise. Quail lay about two hundred eggs a year (almost every day), depending on how much light they have. If you add lighting during the winter, they can produce three hundred or more eggs. Unlike pigeons, quail aren't very good at brooding their own eggs, and like chickens, they need a little extra help. They are more often raised for the eggs than for their meat, because they are smaller than pigeons but do lay more eggs.

The pigeon coop should be at least 6 by 8 feet square (1.8 x 2.5 m), and 7 feet tall (2 m). Each breeding pair also needs a nesting box attached to the wall off the ground, filled with straw or hay, and a constant supply of fresh water. Pigeons

The Ultimate Guide to Urban Farming

▲ Pigeon coop made of recycled materials. The birds can fly out the top.

and quail are social animals, so it is better to have at least three breeding pairs at any one time. There are pigeon breeds just for show, but you are looking for a large meat variety like Cropper, White King, or Silver King. Quail breeds raised for meat include Coturnix (Japanese) and Eastern Bobwhite.

Pigeons and quail can both be allowed to fly free during the day because they will return to the coop at night. They will eat from the garden and fertilize everything, and it is a common practice. There is a much higher chance that predators will

eat them if you do this, but make sure that they get closed in securely at night when they return to the nest. While most coops are designed to walk into, you could also design a coop that sits off the ground like a rabbit hutch to deter possums or weasels.

Breeding

Pigeons mate for life and need very little care. As long as you have a breeding pair together, they will find a nesting box and settle in to make babies. Both take turns sitting on the eggs, which will hatch

---

in eighteen days. Coturnix quail eggs will hatch in eighteen days and Bobwhite in twenty-three days. Pigeons will feed the squabs regurgitated food, and after twenty-eight days they can be butchered, before they start to fly. Quail can be butchered in six weeks. To kill a squab or quail, take it from the nest in the morning before it eats, cut off the head and hang it to bleed out (the same as you would a chicken). The feathers are carefully pulled out without scalding. Remove the feet and throw them out, and cut the body from the vent to the breastbone and remove the organs. Save the gizzard, heart, and liver. Rinse with cold water and refrigerate or freeze as soon as possible. It takes two squabs to make a meal for one person.

It is difficult to tell which birds are male and which are female. You will want to eventually replace your first breeding pairs with younger pairs, so you will have to watch their behavior. Males are noisy and busy strutting, while females are very quiet and sit still.

GEESE

How Geese Fit into the System

Geese eat grass and weeds, and in return will fertilize the soil while leaving your crops and mulch alone. They also protect your property from predators and provide eggs, meat, and feathers. They should only be allowed into a well-established area so that they won't squash any young shoots. They eat fruit and vegetables so they should be removed before your garden ripens.

Geese can be let into the vegetable garden after plants like strawberries and tomatoes have grown to the point that they won't sustain any real damage when the geese walk on them, but usually they will live in their own pen or with ducks. Put

▲ Geese in the garden.

The Ultimate Guide to Urban Farming

seven geese per acre in the field when they are more than eight weeks old, and let them graze until spring when the sprouts come up. There should be a fence around the field that is at least three feet high so they won't get into any other gardens.

Geese can survive with a small pond just like ducks, but heavy breeds won't breed unless they have more water. Six geese are the maximum population per acre of water surface. If there is enough of it, they can also live just off grass pasture and, unlike ducks, they eat older grass. Geese are meat birds, and do well as watchdogs or guards, although they are very quiet. If you keep a goose for a long time, it can become too big for you to handle and become dangerous if you aren't around it every day, but a meat goose doesn't get big enough to pose a threat, because you'll eat it first.

Goose Maintenance

There are breeds of geese for eggs and some for meat. Dark breeds are harder to feather when butchering, which is why you always see pictures of the traditional white goose in farm scenery. It is difficult to tell the difference between a male and a female unless you are practiced in flipping them upside down and looking in their vent. The easiest but less accurate way is to watch the flock for geese with a broader head, longer neck and more aggressiveness. A *gander* (male) has a deeper, louder call and is much more aggressive. It is fairly simple to figure out if you watch during mating. Don't eat a goose more than three years old, which you can tell by the soft, yellow down on its legs.

The goose house needs to be 10 square feet (0.9 square m) per goose, and their yard needs to be very roomy, 30–40

square feet (2.8-3.7 square m) per goose. The house doesn't need to be very fancy, just a simple shed that is very dry. A box feeder can be located inside, but the water trough should be outside under an awning. The floor should be covered with clean bedding such as chopped straw.

There is no commercial goose feed available, but chicken feed can be given if it has 15 percent protein. Geese also need oyster shells or grit at all times. A ration mixture usually includes 10 percent ground corn, 20 percent ground wheat, 10 percent wheat bran, 20 percent ground barley, 21 percent pulverized oats, 8 percent soybean oil meal, 2 percent dried whey, 6 percent dried alfalfa meal, 1 percent ground limestone, 1 percent dicalcium phosphate, and 1 percent iodized salt. If geese have pasture that is not alfalfa, they don't need any extra feed. However, they are able to live entirely off pasture. Although they don't mind eating old grass, they do mind eating alfalfa and generally won't touch it unless they are very hungry. In the winter, provide dried grass, hay (not alfalfa), corn fodder, grain, and whatever scraps you would feed ducks.

Geese need to be clipped just like ducks, after each molting. Clip five inches (12 cm) off the feathers of one wing— careful not to clip the wing itself or during molting, which could cause permanent injury or even death. Geese are big and unwieldy—an irritated goose can bite your face, so always pick them up backwards, with the head facing toward your back and the wings pinned under your arm.

Gather the goose eggs twice a day. If you want to raise goslings, you will need a gander. A big gander can service two or three geese, and a smaller one can handle four or five. Once he picks, he will stick with the same females every year and help raise the goslings. Keep the ganders separate, and let them in with the hens in late fall or early winter. It is recommended to wait until a goose is two years old before breeding because the quality of their eggs is so much better. They will want to brood outside, in an old tire with straw or a tiny brooding house. When she starts to lay, take all of the eggs except two every day until the nest is full. Until she has a full nest, she may not start to set and you could lose some of the goslings anyway, so doing this will also provide you with goose eggs and she will continue laying. It takes between twenty-eight and thirty-five days for the eggs to hatch. Around day twenty, you will have to spray the eggs down completely with warm water and turn them over yourself.

Goslings raised in a brooder need the same floor space as ducks, one and a half square feet (0.14 sq m) until seven weeks and then two and a half square feet (0.23 sq m) after that, but there should only be twenty-five goslings per 250-watt heat lamp. They will need to be fed four times a day, with enough food to eat in fifteen minutes. This includes tender, green grass or weeds, along with a bit of duck food and some grit. In five to six weeks, they will be able to survive solely on a big pasture (1 acre or 0.4 hectare per 20–40 geese). If you don't have a big pasture, you can add some grain. Goslings can be butchered before winter as long as their pinfeathers aren't growing in, which happens in cycles. Since goose tends be greasier than other meats, their grease has traditionally been used for frying, pastry, and hand salves.

Eggs

Gather goose eggs two times a day and clean them like chicken eggs. If you are incubating the eggs, a goose can hatch ten to twelve eggs in twenty-nine to thirty-one days, but they need to be turned by hand three or five times a day if the *setting* goose does not. Never turn them an even number of times or they will lie on the same side every night. You can also incubate in an incubator if you follow the same turning rules. Whether they are incubated under a goose or in an incubator, during the last half (after day 15) of the incubation period sprinkle the eggs with lukewarm water for thirty seconds every day to help with hatching. Remove the goslings from the nest as they hatch. Keep them in a warm place until they are at least three hours old. This prevents the goose from deserting the nest.

Plucking a Live Goose

1. Catch the goose and hold it tightly by both feet. Turn it on its back with its head behind you.
2. Carefully remove only the breast feathers of the goose, without tearing skin or injuring it in any way.
3. If the geese were hatched early, you might be able to pluck them four times a year. A half pound of feathers per goose is a good yield.

Natural Farming and Geese

Geese are natural weeders if you use them before your wanted plants come up. Put seven geese (more than eight weeks old) per acre in the field before the weeds are tall and coarse. A fence around the geese should be three feet high, and let them clear out all the weeds until your sprouts start coming up.

BEES

As a beekeeper you are certain to get stung many times and you can build up immunity. However, you can also suddenly have an allergic reaction. Before buying bees, get tested for allergies to bee stings.

Buy protective bee gear and always work with someone—so that if you do develop an allergy, he or she can get you help. When you get stung, scrape the stinger out quickly with your fingernail so that less venom will enter your skin.

▲ Bees may be the most versatile urban farm stock. These beehives are located in Amsterdam.

How They Fit into the System

Bees are the producers of most of what you eat. Without their pollination, producing enough food to feed us would be impossible. They also make honey and beeswax. The most difficult part of keeping bees is making sure they have enough forage to make enough food to keep them alive through the winter. If they don't have enough, they have to be moved or you will have to add sugar water to the hive to try to keep them alive. In an urban environment, location is key. Bees prefer to fly at least three hundred feet to their food source, and they won't forage well in the face of a cold wind. Using your land map, place the hives away from the wind, and use hedges of herbs to shelter them in the direction that you want them to go.

There are two kinds of forage—pollen and nectar—and bees need both. The pollen species are planted within 100 feet of the hives, and the nectar species are planted at least 300 feet or more away. A line of herb hedges doesn't even have to be more than 3 feet tall, and it directs them from the hive doorway toward the forage by sheltering them from the wind. It can be rosemary, acacia, or built-up soil beds planted with thyme, catmint, or field daisies. You may already have pollen producers around your house that were planted to provide shade: willow, acacia, pine, and vines like grapes. Everything else can be planned to flower in succession so that the bees can have a constant supply throughout the season. Having a minimum of thirty species to forage from is insurance for your hives. These include gooseberries, apples, white clover, blackberries, citrus, buckwheat, mustard, and other fragrant herbs. The rest can be supplied from field crops.

There are three types of honeybees: Italian, Caucasian, and Carniolan. Italians work harder, Caucasians sting less, and Carniolans are the gentlest. Honeybees can only sting once (unlike wasps) because they die, and this makes them less likely to sting. You can either buy bees from a supplier or buy a whole hive from a local beekeeper. The latter option is the easiest because the hive will be well established. Once you have one or two hives, you can have an unlimited supply by encouraging bees to establish new hives.

The Tools

You will need a hive, a smoker, a *hive tool* (a small hooked lever for taking frames out of the hive), bee clothing, a bee brush (for brushing bees off a frame) and a feeder.

▲ **Backyard frame inspection.**

Bee Equipment

It is sometimes better to buy a bee outfit because you know that it will be safe. But to make your own you will need the following:

Very thick leather gloves that are easy to move so you can handle frames easily.

Gauntlets that cover the top of the work gloves and come up above your elbow. They need to be tight around your arm, made of canvas and baggy so the bees can't reach you.

Work boots with your pants tied tightly to the outside of the boots. If bees get in your pants they will crawl up your legs quickly so it is even better to have an additional cover over the top of your boots and pant legs that is tightly sealed.

White, baggy coveralls. If they are made of rip-stop nylon, the bees won't be able to walk on you.

A construction hard hat, jungle-style hat, or bee-proof straw hat with a firm, wide brim. The bee veil is difficult to make, but it goes over the top of the hat and sits on the brim tied firmly with a drawstring. Make sure the drawstring is secure because if your hat comes off then you will get stung on your face. If you do have to use a hard hat or other hat without a back neck protector, you will need to make a cardboard protector, which you can fasten on with duct tape.

You will need a hive, a smoker (although a stick with burning rags might do), a hive tool (for a standard hive), a bee brush, and a feeder. The smoker is the most important; it has a chamber in which you build a small fire and put it out, which creates billowing smoke. A pump pushes smoke out.

A hive has several layers. At the bottom is a *hive stand*, a platform that makes sure the hive is level. Above the stand is the bottom board, a thin frame that holds up the *brood chamber*. The brood chamber is where the bees make their home, and it is where the queen lays eggs, which are deposited in cells to become baby bees. *Supers*, or honey supers, are shallow boxes that sit on the brood chamber and usually hold honey, although sometimes they have baby bee cells. It is better to have a shallow super than a deep one because they can get too full of honey and be difficult to carry. Each super has ten vertical removable frames. On each frame is *foundation*, a flat sheet of beeswax that has hexagons imprinted on it as a template for the bees to build cells on. In most areas used bee equipment is illegal because of disease, so contact your area's department of agriculture before purchasing any.

There is a new invention known as the Flow Hive, which is basically a super that has special frames with partially built honeycomb cells. When you are ready to harvest the honey, the Flow Hive allows you to turn a lever that releases the honey from the cells and pipes it into a jar. At the time of this writing, this invention was so new that it is difficult to give a review, but the professional beekeepers who have tried it highly recommend it, especially for new beekeepers. The price is the same as for a regular hive so there is not much risk here.

Knowing the Hive

Beekeeping is not something that can really be learned from a book. It takes a lot of practice to be able to recognize the different types of cells and bees, which is why your local bee club can help you out (usually for free). The Flow Hive might make the job simpler, but you will still want to learn these basics:

Brood cells—dark-colored caps (unlike honey cells, which are light-colored) and contain baby bees.

Queen cells and bees—one inch long (2.5 cm) and look like a peanut shell that hangs away from the rest of the comb. They contain baby queens. Queen bees are an inch long (2.5 cm) and have a tapered body. They look unique. The other bees won't crowd around them.

Drone cells and bees—stick out like the queen cell, but not as far, and have bullet-shaped tops. They contain baby drones. Drones don't have stingers, are very fat, and have big eyes. Their only job is to compete to mate with the queen. It takes twenty-four days for a drone to hatch.

Worker cells and bees—the smallest cells, are level with the rest of the comb, and contain baby workers. Worker bees are the ones that sting and they keep the hive going. It takes twenty-one days for them to hatch.

Handling Bees

1. In your smoker, start a fire with crumbled paper and add tinder such as pine needles and dry grass. The fuel doesn't need to be too dry because you want it to create smoke. When the fire is burning well, close the lid, and use the pump to keep it smoldering.
2. Stand to one side of the entrance to the hive and blow smoke in the door. Wait a minute or two, take off the cover and blow more smoke in the top.
3. Whenever the bees start to get agitated with you, use more smoke. Be careful not to hurt a bee or it will release a panic odor alerting the other bees to sting you.

The hive can be inspected if the temperature is over 50°F (10°C) and the weather is nice. Some people take a look once a week but that's a bit more than the bees are comfortable with. In general, it is only necessary to check in if you suspect a problem, and in the spring and fall. In the spring, carefully remove every single frame and find the queen. Look for queen cells, find out how many bees there are, how many brood cells there are and what type, how much honey is coming, and whether they need more supers. Supers prevent overcrowding, which prevents swarming.

Bees always need fresh, clean water. However, big ponds don't make a good water source because the bees may drown or be killed by a resident dragonfly. One or two hives only need a nearby outside faucet left to drip onto a slanted board. If you have lots of hives, a soaked mat near the pond or a very tiny pond near the hive can supply them with water. If you see bees standing around outside the door of the hive in warm weather, and you know the hive has a high population, it means they are having trouble cooling the hive. Move the hive into the shade, make the entrance larger, and stagger the supers for ventilation. In the winter, making the door much smaller will help them keep heat in and also prevent mice from creeping in and stealing honey.

Bees fare better in a warmer climate—when they are less likely to freeze—because of the greater availability of food. Placement is the key to a successful hive. Point the door of the hive in the direction you want them to go (usually toward the pollen-producing plants), and away from houses and barns and loud motors. Every hive will need fifty to one hundred pounds of honey to get through the winter. If they get low on feed them 2 parts granulated sugar per 1 part water, and in the spring you can give them artificial pollen.

Making New Hives

To move an entire hive, plug the door of the hive very tightly with a porous material that allows air in, so the bees won't suffocate. It must be very secure or you will find yourself in the middle of an angry swarm. To create more hives, start at the beginning of May. You can take four frames with brood cells from your most established hive, as well as some honey and *bee bread* (or pollen). Bee bread is yellow and grainy. You will also need worker bees, which you can just brush into the hive. It is better to have one queen cell per frame also, but if there is not, they will make one. Put the frames into the new hive and stuff the door loosely with grass. It can take two weeks for the hive to produce a queen, and then the queen needs four weeks to mature and mate. The bees will put so much effort into this process that they will only make enough honey to support themselves over the winter, and you will not be able to collect any honey that year from the source hive or the new hive for yourself. You also run the risk of losing a new colony if they fail to feed royal jelly to the queen on the first day of hatching.

Keeping Bees Healthy

Keeping bees healthy is your most important job. If you notice a sick bee, you have a responsibility to inform local bee inspectors so that you can prevent a *die off.* This is when a bee population in a geographic region is decimated because bee disease spread quickly. Follow these good practices:

Only buy bees from a place with a good reputation.

Don't buy used equipment unless you've talked to your local apiary official.

Replace foundations every two to four years.

Watch for disease and infestations every time you open the hive:

- bees that can't fly
- discolored or misshapen cells
- punctured or sunken cells
- dead larvae
- dead bees
- swollen bees with shaking wings
- mites with eight legs
- gray webs in the comb

Beekeeping calendar:

Early spring: Check that they have enough food, and supply them with artificial pollen. In some cold climates, you might see dead bees at the bottom of the hive, but usually this means the queen has died. If the queen dies and there are no eggs, the worker bees will wander around and eventually die, too. If they do have eggs or larvae they will make a new queen. The workers should keep the hive very clean. If it gets dirty, it's a sign that the queen is gone and they will all die.

Late spring/early summer: When you think the bee population is big enough, add another section to the hive to sustain the increase in comb production. This is also the time to split the hive into two hives if you want. If the bees feel too crowded they will swarm—that is, they will leave the hive as a group. They won't sting, but you may have to track them down and coax them into the hive. Splitting the hive and adding sections prevents this.

Fall: On a warm sunny afternoon, take out the honey. Leave at least fifty to one hundred pounds for them to eat during the winter, depending on how long your winter is.

Winter: Keep the hive very well ventilated, and protect it from wind. Check their food supply and add sugar or sugar water to keep them from starving.

Types of Bees

Honeybees: They can only sting once because they leave their stinger in your skin and then die. This makes them more careful with their stinger and less likely to get you. There are three breeds: Italian, Caucasian, and Carniolan. The best way to get bees is to order them from a local company or purchase a hive from a local beekeeper. Some groups have been breeding bees for microclimates in the hope that they will be more resilient to disease. As mentioned earlier, bee diseases spread very quickly, but can be prevented through better genetics by raising bees that are can resist the types of diseases found in one area.

Wasps/yellow jackets/hornets: They don't eat flower nectar, they eat bugs, fruit and other foods. They are great to have in a garden because they eat thousands of pests in one season, but they will also sting for any reason, and over and over again.

African bees: There is a lot of hype about these bees since they escaped from a lab in Sao Paulo, Brazil. They tend to attack as a swarm and have killed a few people. You can outrun them if you can run fifteen miles per hour for five to seven minutes, or eight to ten city blocks. Don't ever use bug spray because it just makes them angry—the only way to deter them is with huge quantities of soapy water, from a fire truck.

Types of diseases and pests

Acarine: These tracheal mites will bore holes in the air passages of bees one to eight days old and suck out their blood. Keep the bees producing babies so your population goes up despite the deaths. You may need to use a Terramycin treatment.

American foulbrood: This is the worst brood disease, a bacteria that causes larvae to rot. The cell caps will be an off color, sunken in, and punctured. The larvae will be dark brown, slimy, and smell like rotten eggs. By law, you must report it to the state or provincial regulator. To prevent it, you can use Terramycin powder as a treatment once in fall and once in spring. It can develop when your bees can't remove dead cells fast enough. A healthy, balanced colony may be able fight the disease successfully. You can tell by watching a few minutes and see if any dead stuff is being taken out of the hive.

Ants: If you have ants in your hive, they are a symptom of other problems. If the colony has become weakened because of disease or a failing queen (which may cause the workers to pull back from the brood), ants will plunder the colony. Sometimes this can make the bees leave.

Chalkbrood: A fungus that causes the larvae to die. They turn from white to gray, to black, and then get hard and chalky. Take out the infected combs and burn them.

Chilled brood: Not a disease but the brood is too big for the outer edge to get warm in cold temperatures. You will find

The Ultimate Guide to Urban Farming

bees dead at all stages, from babies to adults.

European foulbrood: This disease resembles American foulbrood but there is no treatment. Only an experienced bee person can tell the difference by pulling out the larvae with a stick—it won't be stringy.

Nosema: Bees naturally carry a parasite in their intestines called Nosema apis. If they don't get enough pollen, the nosema multiplies and kills them. Bees will be swollen and crawl around outside the hive with their wings shaking. Make sure bees have enough pollen, and there is a preventative medicine called Fumidil-D.

Sacbrood: A rare virus that makes cell caps look dark and sunken. Larvae will look gray and black.

Stonebrood: Very rare fungus that makes cells green and mildewy. The bees fight this disease themselves.

Varroatosis: A mite from India that will kill European bees. It is light brown, oval, and has eight legs (if you see six, it is a harmless louse). It kills drone cells. Use plastic pesticide strips or the colony will die.

Wax moths: They lay eggs on the combs and then the caterpillars eat through it when they hatch. You will see fine gray webs around paths and tunnels through the comb. If your colony is strong, they will kill them.

Removing honey and making wax:

1. Some suggest heating wax in a double boiler in order to purify the honey and remove the wax. This not only ruins the color, it also drives out the oils and fragrances destroying the flavor. The best way is to hang the honey in a strainer bag and allow gravity to do a perfect job.

2. Take the wax and put it in a box and set it on top of the hive. Leave it one day and then remove it—the bees will have cleaned all the honey off it.

3. Wrap the wax in a thick cloth such as sweatshirt material. Put the wax in a double boiler on low heat and melt it. As it drips through the cloth it will purify.

4. If the wax looks dirty, and you have a woodstove, add a little cider vinegar and a little water and keep it at 135–140°F for two to three days. The dirt will settle to the bottom and honey will sit just below the wax floating on top.

▲ Many bee clubs own a centrifuge for extracting honey. If you don't have this, gravity does the trick.

GOATS

How They Fit into the System

Goats are exceptionally good at clearing pasture and effectively clear the toughest brambles and unwanted vegetation. They can be temporarily used for this purpose by penning them or tying them with a halter and moving them from place to place. Goats can be so destructive, however, that it is only recommended to keep a few for milk and meat production. More than one goat per person in your family is unnecessary. If you do use lactating goats to clear a pasture, you will have to give them a little bit of grain to keep their milk production up.

Goats aren't necessarily allowed in the city. Things are changing in many cities by people who already own goats and are willing to face the music if they get found out. The key to success here is making sure your goats look and smell great.

How Many Goats

An average doe makes about three quarts of milk per day: one to one and a half quarts goes to her kid, leaving only one and a half quarts per day. If she is a new mom, she will give even less than that. The more does you have the better, because they can feed their own babies. They also adopt orphaned babies of other animals, feed extra milk to other animals (such as chickens), make cream and butter, and supply you milk. Have one goat per person in your family.

Before you buy a goat, find out what breed she is, the age, if she has been bred and to what kind of goat, and how many times. Find out how many kids she had if she has been bred, how long she milked, what it is like to milk her, how much milk

▲ A clandestine goat.

she gives, how many teats she has, and if she jumps fences. Find out if she is hard to catch, if she bites or butts. Has she had a distemper shot, and will she wear a halter? See her parents, sisters, or brothers, or other family members. If you buy a buck, remember that they stink, and they must be handled carefully. You will need to decide whether to keep it separate or with the herd (see below).

Goat Management

Goats are clever and can also jump high. The fence should be at least four and a half feet (1.4 m) high, with a quarter acre (0.1 ha) per goat. Wrap trees with chicken wire so they can't strip the bark off, and make sure the fence does not have a gap wider than 8 inches (20 cm). If the goats can't see through it, they won't try to get out. If you must have a rail fence, make sure they can't squeeze through. Goats can unlock most standard latches with their tongue so a padlock may be necessary. If the goat does try to get out all the time, put a Y-shaped yoke on its head so it can't fit, and soon it will give up trying. Then you can remove the yoke. The goat house can be any kind of sturdy three-sided shed in their pasture, or in the barn as long as there is 36 square feet (3.3 sq m) per goat, with clean hay for bedding.

Goats are very curious, so pretend to examine something while showing her some grain. Not only will she want the grain, she will want to see what you are looking at. You will need a second person to help you, and when the goat walks closer to investigate, the helper can grab her. If you spend a lot of time with your goats, this may not be necessary, as they can become quite attached to you. Here's how to find out the sex of the goat: a boy pees from the middle of his belly, and a girl has to pee squatting backwards a bit.

Each goat needs about four to five pounds (1.8 to 2.3 kg) of hay per day of mixed grass and legumes, such as alfalfa. That's about half a bale of hay per ten goats, twice a day, placed on a hayrack. Since you probably have them on an overgrown pasture, they will eat less hay. Goats also need salt and water at all times. Goats won't lick a salt block, so you will have to provide loose mineralized salt. While goats are destructive and curious and tend to get out of their fencing, in other ways they are very easy to care for. If they get lice, rub them down with vinegar. Trim the hooves once a month using a knife or

> **Goats will try to eat anything, including poisonous plants. Check their pasture for these common dangers:**
> Milkweed
> Nightshade plants
> Buckthorn
> Cowbane
> Dog's mercury
> Foxglove
> Greater celandine
> Hemlock trees
> Henbane
> Ragwort
> Rhododendron
> Rhubarb leaves
> Spindle
> Water dropwort
> Yew
> Iris
> Azalea
> Beet leaves
> Evergreen trees

hoof nippers, or they will keep growing. Goats are very susceptible to worms and under regular standards would be wormed three times a year. Under organic certification worming is not allowed. The goats themselves will control their own worms in the way that they eat. They eat higher leaves first and work their way down, and wander far distances over their pasture, which is why the tree method of the pasture is particularly valuable for them. They should be rotated to new pasture every three weeks, which is the lifespan of a stomach worm. They need to be kept in very clean conditions with fresh, clean water readily available at all times. Unless you have lots of room for the goats to roam and lots of forage available off the ground, you may have to worm them.

Breeding

Breeding does need a quarter pound of grain per day starting on October 1st and increasing a quarter pound (0.11 kg) per week until the start of November when each doe is getting one pound (0.5 kg) of grain per day. This will increase the chances of having twins and triplets. For healthy older does, taper off this grain starting December 15 (six weeks after breeding), and then bring it back up again starting February 15 (six weeks before kidding). First-time breeders and unhealthy animals can keep eating grain all the way through. Bucks that are running in with the does will have access to grain as well, but it should be gradually tapered off when they are in good condition.

Check the mother's udder twice a day after kidding to make sure it isn't too full. Her milk will come in three to five days after kidding, so during that time don't give her grain or it will come in too fast. What you

feed your goat can change how the milk tastes. If the milk tastes bitter, try removing high-odor foods like garlic or cabbage from her diet. Goat milk may taste naturally "goaty," but you can fix this by putting a pan of baking soda in her feed trough. Keep it full of soda and in a few days the milk will taste sweet, and in fact any goat eating grain should eat baking soda to avoid a sickness and too much acid. Pregnant does who are producing only a little milk should stop milking two months before kidding, but you can keep heavy-producing does milking all the way through continuously. Pregnant milkers need high-quality feed, but you should still watch their weight.

To have milk from a doe, she will have to breed with a buck. You will not be able to keep a buck in the city, not only because they can be very aggressive but also because they really stink. You will have to rent a buck or purchase semen.

Breed your does 149 days or five months before you want to have kids. For small farms, having kids around April 1st is ideal, so breed on November 1st. Don't breed does that are less than seventy pounds or two years old because they will have health problems. The doe will almost always kid exactly 149 days later. Does are in heat when they spend time sniffing and wagging tails toward the buck pen, and they will make more noise.

When a doe is close to kidding, you should check on her every morning, and if she has delivered you will need to find the kids as soon as possible. A barn floor with dry bedding is ideal if they are familiar with the place. Give her clean, fresh water in a small bucket and lots of good hay. When she starts to give birth, you can clean the nose and mouth to help the kid breath. Otherwise, leave her alone unless

> **Ketosis**
>
> Pregnant does can be susceptible to ketosis, with symptoms that include dullness, lack of appetite, grinding teeth, and wandering. To prevent this, give the does blackstrap molasses during the last two months of pregnancy. Stress, overfeeding, underfeeding, or lack of exercise can cause it. It's a good idea to trim the hooves at this time as well.

- her water breaks and two hours pass without seeing any part of a kid
- she's in great pain and thirty minutes pass with nothing happening
- she's totally exhausted and fifteen minutes pass with nothing happening.

To assist in the birth of a head-first kid, pull gently downward with each contraction. If one of the situations above occurs, or you can see a bum or neck coming out first, scrub your hands and arms well and make sure your nails are very clean. Reach in very carefully, determine the kid's position and gently reposition it until its feet and nose are coming out together. After delivery, give the doe a bucket of warm water with some molasses in it and let her deliver the afterbirth. She may eat it, or you can bury it.

Wipe the kid's face, and if the goat isn't doing a good job of drying the kid, then go ahead and dry it off. If you find a kid outside after birth, bring it inside and wrap it up next to a heat source until it is warm. It isn't necessary to tie the cord, just dip the end in iodine or alcohol to kill bacteria. It's common for a goat to have two or three kids, and they should be standing up right away. If one doesn't nurse within the first fifteen minutes, help it by holding the teat in its mouth so it can suck. If it doesn't suck, squirt milk in its mouth and then try to get it to suck again in three or four hours, or when it seems hungry. Some kids need help like this for three days. If it still doesn't suck after several hours, then you'll have to bottle feed, but only a couple of times so it learns to suck—then try to get it back on the mother. Don't keep the kid in the house or away from the mother for more than six hours or the mother will reject it. For three to five days after birth, the kids and mother should be kept separate from the herd until they are strong and nursing well.

A rejected kid must still be fed colostrum. Give the mother a little grain and milk her. Save the colostrum and put it in a bottle with a lamb nipple on the end and feed it to the baby. If the kid is too weak to use the lamb nipple, use a human baby's nasal aspirator (or syringe). Never feed a kid cold milk. It is so important to get the colostrum into the baby, otherwise it could die. The kid will have to be fed two ounces (57 g) every two hours on day one, gradually increasing to three ounces (85 g) every three hours on day three, and six ounces (170 g) every four hours on day seven. Work your way up to eight ounces (226 g) morning, noon, and night for two weeks, and then gradually to sixteen ounces (0.5 k) morning and night after you milk. If you can't give the kid goat's milk, the next best is cow's milk, but they can still get scours (or diarrhea) and die more easily. You will have to butcher the kid if you don't have goat milk, or if you need the milk yourself. It is possible to overfeed a kid. Normally a mother goat lets the kid start eating and then walks off after less than a minute. Frequent, small meals are better. After they are two or three months old they can be weaned, but a bottle-fed kid will be more attached to you and become annoying.

Let the kids stay with their mother until they are strong and eating solid food well. This usually takes two months for singles and twins, and three months for triplets. Then separate the kids from the mother at night and milk her in the morning, then put her back with the kids. In a few days they will learn the routine, although they will make a lot of noise about it. As the kids wean, keep milking her more and more so she will get used to holding more milk in her udder at a time and her teats will lengthen. You can either keep letting the mother feed the kids at night and only milk once a day, or you can gradually work toward milking her every twelve hours. If you let the mother nurse once a day, she will eventually wean them herself.

Besides the rejected kids that you can't feed, it is also likely that you will end up with more billy goats than you need. Nanny goats are valuable and you will hardly ever need to kill them because you can sell them instead, but billies don't give milk and if you have too many then you get no milk for yourself. All of the extras can be raised for meat until the fall. If you have them butchered in November, you won't have to feed them over the winter.

Milking

Stainless steel equipment without seams are the best and also the most expensive containers for milk. Food grade plastic and glass will work if they are seamless, but don't reuse plastic milk jugs from the store. You'll have to purchase containers. Rinse buckets, containers, and utensils in lukewarm water right after you use them or they will collect difficult-to-remove milk deposits. Wash them after every use by scrubbing thoroughly in warm, soapy water, then rinse again in scalding water. Air dry upside down. Any cloth used for straining should be rinsed and boiled directly after.

▲ A goat-milking stanchion.

Goat milking troubleshooting:

The milk won't let down: Massage the udder either with the cleaning cloth or with bag balm, or gently pat the udder like a kid butting her. If she still won't let down, real goatherds suck on the teat a little.

Drying up milk: This is done to let her have another kid. When you milk, leave a little milk in it. When her milk production has reduced, milk her only once a day. Be aware of the signs of mastitis as this is when she is most at risk.

Mastitis: The first sign is milk with strange textures: flakes, lumps or strings. Don't drink it and don't throw it somewhere that an animal can lick it! Wash your hands very well after touching the animal because it is infectious. Even healthy does should be checked at least once a week by squirting the milk into a cloth. Feel the udder for tumors, large hard areas, or an *abscess*. An abscess is a red, tender swelling of a whole side of the udder, which makes it warmer and more difficult to get milk from. If you allow it to get worse, the milk may turn yellow or even brown or pink from pus and blood. The only cure is antibiotics. A bruised udder, getting too full for too long, or having had a previous case of mastitis makes her more susceptible.

Self-sucking: Goats are flexible enough to suck their own milk. Other than butchering, you can use an Elizabethan collar or side-stick harness that can prevent her from reaching the udder.

Getting the milk:

1. Clean the milking utensils in warm, soapy water.
2. Put the goat in a *stanchion*, a frame that holds the goat by its neck. Many people have a milking stand that raises the goat a little higher for milking comfort. Brush the fur to get out loose hair and dirt, put down fresh bedding and keep the long hair under the udder clipped.
3. Put some feed in the stanchion's trough. Milking is done every twelve hours, starting early morning before they eat. Always be on time or the goat will get too full, which is painful. If you aren't a morning person, just make sure that you really stick to a schedule.
4. Brush the doe and look it over for problems. Then wash your hands and dry them, and fill a bucket with water 120–130°F (49–54°C).
5. Wash her udder and teats. This helps the milk let down, and removes dirt and bacteria, making better milk and a healthier goat. Wait a minute after washing to start milking.
6. The goat needs to be happy and relaxed for the milk to let down. Put your thumb and forefinger around the teat near the top of the udder, pushing up slightly and allow the teat to fill with milk. Then close your hand around it and squeeze the milk out while pulling down. Keep your hand away from the nipple hole or the milk will go all over. Squirt the first three squeezes into the ground because the first milk has more bacteria in it. Make sure you completely empty the udder or she will produce less and less until it's gone.
7. Strain the milk. Use a regular kitchen strainer lined with several layers of clean fabric such as a dishcloth, muslin, or even a cloth diaper.
8. If you choose to pasteurize, get a double boiler. Use a thermometer to heat the milk to 161°F (72°C), stirring constantly. Maintain that temperature for twenty seconds, then quickly remove it from heat and immerse the pot in very

cold water, stirring constantly until it gets down to 60°F (16°C).

9. If you don't pasteurize, put the storage container inside another container full of cold water. It needs to be chilled to 40°F (4°C) within one hour. Store any milk in the coldest part of the fridge.

Besides baking soda, to prevent the "goaty" taste in milk, you can prevent a distasteful flavor in milk by only using seamless stainless steel, food grade plastic, or glass to store milk. Keep milk out of sunlight and fluorescent light, keep the barn clean, and clean the udder and your hands. Don't feed the goat strong-flavored foods (such as onions, garlic, cabbage, or turnips) less than seven hours before milking, and don't let her smell the buck or let the buck near the milk. Don't smoke around the milk.

Processing Goat Milk

Goat cream is harder to remove than cow cream. Cow cream will just rise to the top if you let it sit. If you put goat milk out for twenty-four hours some of the cream will rise but not all of it. To get it all, you will need a cream separator specially made for goat's milk, or you will have to put the milk in the fridge for five days and skim off the cream. Run the milk through the separator and save the cream. The rest is your skimmed milk. You can freeze cream for later if you thaw it completely before using.

Canned milk:

Pour the fresh milk into clean jars up to ½-inch from the top of the jar. Put on the lids and process in a pressure cooker for 10 minutes at 10 pounds of pressure or 60 minutes in a water bath canner.

Make curds and whey:

Let the milk sit out in a covered enamel roaster, a canning kettle, heavy crockery or a stainless steel bowl until it gets completely sour or "clabbered." Don't use aluminum! Keep the temperature at 75–85°F. It should be placed on a warming shelf on your wood stove or other warm place as if you were letting yeast bread rise. However, it should not be *too* hot. When the curds separate from the whey, it has "set up." It will feel like jelly, and it will form a single large curd floating in the whey. If you need clabbered milk sooner, add 1 tablespoon vinegar or lemon juice per cup of milk. Goat milk takes longer to set up, as many as five days.

Churning cream into butter:

1. You will need some kind of churn to make butter. Either use a traditional churn or make your own using a quart jar with a lid. Again, people have used an electric mixer with success.

2. The cream must be 60°F, or about room temperature, and put it in the churn or jar so it is half full.

3. Churn steadily and rhythmically, not too fast and not too slow (if you are using a jar, simply slosh the cream back and forth). After twenty to thirty minutes butter clumps will form.

4. Remove the butter from the milk (the milk is buttermilk) and put it in a bowl of cold water. Roll it into a ball, then flatten it and repeat just like kneading bread. Change the water frequently when it gets cloudy, and as the milk is rinsed away the butter will start to feel waxy.

5. Keep working the butter until the water stays clear, then take it out and squeeze out the water. Mix in flavoring (such as honey, salt, etc.) if you want, then pat it dry with a clean towel and press it into a mold or form it into a cake shape. Cover and put in the fridge or freeze for later.

Making yogurt:

1. Use pasteurized milk so the good bacteria don't have to fight bad bacteria. You can make yogurt right after pasteurizing if you cool it to 110°F (no hotter or it will kill the acidophilus bacteria).

2. While the milk is still on the heat, add 1 heaping teaspoon of cultured starter per quart of milk. Stir gently, then remove from heat and pour into clean warm jars.

3. Put the milk in a warm place and keep it warm. A few easy ways to do this are put the jars on a heating pad and cover with a towel, or place them in an insulated cooler filled with 100°F water, or use your oven.

4. Once it's warm don't disturb it. Six to eight hours later it should have thickened and then it is ready. Add a half cup powdered milk to it if you want it to be thicker, and mix in flavors such as fruits, honey, or vanilla. Refrigerate immediately.

Yogurt starter:

Most people use plain yogurt from the store that has live cultures. Make sure it is live—some yogurt is not. Make sure it is well stirred, and keep it separate from the starter you will be using. Save one-half cup of your first batch and use it as your next starter within five days. If your yogurt gets a weird taste or smell, scrap your starter and start again.

SHEEP

How They Fit into the System

Sheep can be let into the orchard to forage after the trees are at least seven years old, but they must be carefully controlled. They must be taken out if they start to damage the trees. However,

▲ City grazing as an alternative to mowing.

just because the orchard is not available doesn't mean trees can't be developed as a mainstay of their food within their grazing area. Trees provide food and protection from the elements, benefit the soil, and prevent erosion. Sheep also need a salt lick and a variety of other food types. A legume-grass mix pasture will feed five ewes and eight lambs per acre in a northern region, but it needs to be rotated every week. In the last month of pregnancy, ewes need a half to one pound (0.22 to 0.5 kg) of grain per day. After lambing, sheep with one lamb need one pound (0.5 kg) of grain; sheep with two lambs need one and a half to two pounds (0.7–0.9 kg). This can be gradually tapered off in the next two months. A ram needs one pound of grain each day during breeding season. This grain supplement can be offset by tree forage. Each sheep also needs one and a half gallons (6 L) of fresh, clean water every day. Every spring you will need to gradually introduce the sheep to pasture so they can adjust. In the winter they can forage from a harvested corn field and roots left in the garden. Each sheep will eat seventy-five pounds (34 kg) of grain and ten bales of hay over the winter.

Sheep Management

Sheep don't need much housing. A three-sided shelter in the pasture is sufficient, with a place in the barn for lambing or extreme weather. Sheep are very vulnerable to stress. Moving them in a truck, abrupt feed changes, or even loud storms are enough to make them stop eating and even have a heart attack. They are also an easy target for predators. A dog is the most effective prevention. A border collie will herd the sheep, or you can have a guard dog live with the sheep in the field.

Breeding

Breed your sheep in October and in five months you will have lambs. If your sheep have hair on their faces, clip the wool away from the eyes, and also *tag* her, or clip the wool away from the vagina. Every time you do something to a ewe is a good time to check her feet. You can buy a yearling ram every year from an unrelated herd to breed with the ewes and then eat him afterwards, which is probably your best solution in an urban environment. Obviously it's a bad idea to use rams from the same family for breeding. Lambing is almost exactly the same as kidding for goats, and orphan lambs can drink goat milk as a replacement.

What to look for when buying sheep:

- Ram: Has good legs and feet. Does his job as a ram.
- Ewe: Doesn't have mastitis. Udder is soft and in working condition.
- Is not overweight or underweight.
- Does not limp.
- No hooves are hotter than the others, and no green tinge (signs of foot rot).
- Mouth is sound, gums have no anemia and are very red.
- Eyes have no anemia, arteries around eyes are red.
- No swelling or lumps under the chin.
- Ask about sheep history.
- If possible, ask if the sheep had rectal or uterine prolapse (lining protrudes from body); it is genetic.
- A sheep with a "broken mouth" (teeth missing and crooked) is not worth it.

Sheep Pasture and Feeding

A good pasture will feed five ewes and eight lambs per acre for the season in northern regions. It's a good idea to have 25–50 percent more than that for management. Legume-grass mixtures are good for sheep. Alfalfa-grass and trefoil-grass will regrow in fifteen to twenty days. Rotate their pasture every week. Put loose salt (not in blocks) under cover outside for them. Sheep need one and a half gallons water a day, fresh and clean. Make sure that you clean the feeders before giving food, watch to make sure that each sheep is eating well, and every ewe should have one and a half feet of rack space. If you are using the wool, you may need to keep the wool clean—don't let burdock, thistles, fleabane, Spanish nettle, or other burred plants grow in the pasture where they will rub on the sheep (give them to chickens). The pasture should be divided in half so that you can rotate when the grass is eaten down. Rotation helps prevent worms. The following is a list of feeding guidelines:

Pregnant ewes: Feed ewes in the first four months of pregnancy three and a half to four and a half pounds alfalfa or clover hay. In the last month of pregnancy give one-half to one pound grain also, such as whole oats or shelled corn.

Ewes after lambing: Give one pound of grain per head to those with one lamb, one and a half to two pounds to ones with twins. After two months, feed one half to one pound grain. When lambs are weaned, take away all grain suddenly, and only give a little hay for the week to help dry up the ewes.

Before breeding: During breeding season give rams one pound of grain a day. Put ewes on top quality hay, lush pasture, or give some grain.

When turning out to pasture: Every spring when you turn out the sheep to pasture wait until the grass is a few inches high and give them a big pile of hay so that their feed will change more gradually.

During winter: You can let your sheep eat from a harvested corn field, give them roots from the garden, and two to four pounds of hay per day per sheep, or ten bales of hay per winter. Plus give them seventy-five pounds of grain.

Creep feeding young lambs:

Start feeding lambs at two weeks even if they will be finished at pasture. Grind the feed at first, then feed whole. The creep feed should contain 15 percent protein, and 20–40 percent high-quality alfalfa hay given in a separate rack. A lamb will typically eat one and a half to two pounds of rations per day from 10 to 120 days old.

Sheep Diseases

Take particular care of sheep after stress, such as storms, transportation, or abrupt feed changes. They may stop eating, or even have heart attacks. Symptoms of diseases include breathing, dullness and listlessness, refusing to eat, lying down a lot, and going off alone. Keep a rectal thermometer and if you see any of those symptoms, check their temperature. Normal temperature is from 100.9 to 103.8°F, average is 102.3°F.

Enterotoxemia in lambs: Symptoms include tremors and convulsions, diarrhea, and collapse, and death usually occurs in two hours. It is caused by the bacteria *Clostridium perfringens.* Lambs get it from overeating and quick transition to rich foods. To prevent it, make sure to transition to new foods gradually, having a scheduled and fixed feeding regimen and vaccinate pregnant ewes. It cannot be cured once caught. It is prevented with *Clostridium perfringens,* Type C and D toxoid, administered by a vet during pregnancy and a booster at least four weeks before lambing.

Vibriosis: Causes abortions, dead and weak lambs. It is the most common cause of sheep abortions. Ewes and lambs require two vaccinations, then yearly boosters.

EAE: Caused by *Chlamydia psittaci.* Produces abortions, eye infections, lamb arthritis, epididymitis (fertility problems in males), pneumonia, and diarrhea. The vaccine is usually given with vibriosis bacterin.

Leptospirosis: Causes abortions, anemia, and systemic disease. A killed bacterium is given to immunize sheep.

Tetanus (Lockjaw): Clostridium tetani enters the body through open wounds and causes muscle spasms, stiffness, and other nervous system problems. Seventy-five percent of lambs infected from tetanus die from it. A Tetanus toxoid vaccine every ten years prevents it.

Clostridial disease: Also called blackleg, malignant edema, and braxy, it is a soil bacterium found in the intestinal tract. Symptoms are lameness and swelling just beneath the skin (subcutaneous), and rapid death.

Bluetongue: Also called sore muzzle, the virus is spread by the biting midge. Produces mouth ulcers, nasal discharge, crusty nostrils, and lameness. The vaccine is a live virus vaccine that must be introduced to a flock only if they have it because it will give it to them. Wear rubber gloves and follow directions.

Sore Mouth: Also called Contagious ecthyma, causes pox, or lesions on lips, nostrils, eyelids, mouth, teats, feet, etc. Spread through direct and indirect contact.

Predators

Sheep are especially vulnerable to predators since they are not that smart and get scared very easily. Sheep have died from heart attacks when hearing thunder. Dogs can chase a sheep to death without biting it, and of course any larger animal can and will eat a sheep. The prevention to this is a dog. Either get a border collie to herd the sheep when you need to (they also are great at watching children), or get a guard dog to live with the sheep in the field. One of the best guard dogs for this purpose is a Great Pyrenees. They stay in the field and attack sheep predators. To raise one of these the ideal way is to have a ewe nurse and raise the puppy and it will become part of the sheep family but keep its guarding instincts.

Breeding

If the sheep is too fat, start a diet six weeks before breeding or she will be at risk for pregnancy problems. Two weeks before breeding, "flush" the ewes by putting them on the richest pasture or giving them really good alfalfa hay, or give them 50 percent more grain. This will increase the chances of having twins or triplets. If the sheep breed is not "open-faced" (it has hair on its face), clip the wool away from the eyes, and also "tag" her, or clip the wool away from the vagina. Check her feet and trim at the same time. It is common to schedule lambing for February, so since sheep-mating season is from September through December, breed in October. In five months, she will lamb.

Get your ewe into a lambing pen that is completely dry, protected from wind, with a room. Check your herd every two hours because the sheep will give birth all around the same time. Gather together iodine, warm water, soap, lubricant, and old clean towels or blankets.

Lambing

Note: Make sure there is a veterinarian or knowledgeable person available around the time of lambing if possible.

1. If the ewe's water broke and she has been straining for twenty to thirty minutes and no lamb is born, you will need to help.
2. If you haven't tagged her already, clip the dirty and excess wool from around the ewe's birth canal. Then wash her with warm, soapy water.
3. Wash your arms with warm, soapy water and apply an antiseptic lubricant. You might want to use a sterile obstetrical sleeve instead.
4. Enter the ewe and feel the lamb to see what's going on. Find out how many there are, and which feet go with which legs before you start pulling.
5. Deliver the closest lamb first (if it's twins). Pull gently downward between the legs of the ewe, rotating gently if you need to free a shoulder.

What to do if it's a breech:

If you feel a tail instead of a head, call a vet. If you can't get the vet, get an assistant. Have the assistant to hold the ewe up by the hind legs. Put on a rubber glove and push the lamb forward. Then pull its hind legs straight (they will be bent forward). Gently but quickly pull the lamb out the same as if it were head forward. Clear out its nostrils.

New lamb care:

Pinch or cut off the umbilical cord four to five inches from its belly. Dip the navel into iodine. If the lamb is really chilly, wrap it in a blanket or towel and bring it into the

---

Milking

Sheep's milk makes excellent cheese and makes more cheese per pound than cow's milk. It is easier to make into cheese than goat's milk also. You should get 1 pint–1 quart of milk per day. It also makes great yogurt and ice cream but not very good butter because it is naturally homogenized.

---

house to get warm. Then get it back to the mother quickly to start nursing. It should nurse within thirty minutes of birth, and if it isn't interested, milk some colostrum into its mouth. If the mother isn't doing what she's supposed to, rub some placenta on the baby, especially around its anus. Even an orphaned lamb can be adopted by a ewe if you rub a new placenta on it.

Bottle-fed lambs are called bum lambs. If you have goat milk, use it. If not, buy a replacement with 30 percent fat, 20–24 percent protein, and less than 25 percent lactose. Orphan lambs occur somewhat frequently because a ewe will not recognize its lamb if it does not smell right. If you take the lamb in the house to get warm and bring it back, the mother may have totally forgotten it. Feed every four to six hours until it is three weeks old, then every eight hours for the next two months. The first few feedings the lamb will not drink more than 2 ounces.

Shearing

Shear during dry weather. Put the sheep in a small pen or in a barn with a clean floor or tarp nearby. Have someone hold the sheep but not by the wool. Put the sheep between your knees on its side with its head pointing toward your back. Cut off any dung locks (wool with poop

---

it for dirt, let it dry, and gather it up with the side that was against the sheep's skin on the outside. Don't store it with plastic, instead use paper or cardboard. When you are done, go have a medical professional check you for ticks.

AQUACULTURE

Aquaculture isn't just a pond. It is a closed-loop system of growing fish in a cycle with plants and other animals. These range in size from small backyard ponds to large intensive aboveground tanks. Historically, the South American chinampa structure of canals stretching between large planting beds were filled with fish and were some of the earliest examples of highly productive aquaculture. Because aquaculture does have such high production value with relatively little effort, it is very tempting to set up the large tank

in it), grease tags (knots of matted wool) and throw them out. Then clip the wool off, cutting as close to the body as you can (about ½ inch away), and it should come off all in one piece. Start with the head, clip across the body from one side, across the back and to the other side, zigzagging all the way to the rear. Don't cut the skin, and don't cut twice—if you don't get it with the first clip, forget it. Treat any cuts with disinfectant, check the hooves, and let her loose. Lay the fleece out and examine

▲ A closed-loop aquaculture system.

The Ultimate Guide to Urban Farming

system and turn it into intensive agriculture. Some very profitable methods have appeared in which one species tank-grown fish are grown with plants in an aquaponic system. The value of the diversity of species is its ability to become a self-reliant and self-contained system that uses very little effort and grows a variety of food.

If you have set up a multiple-pond system, with several small parallel ponds for younger fish, or *fry*, you will have a much easier time managing your fish stock. Fish will breed and your fry will get eaten unless you segregate them. Raise the fry in one of the small ponds and release them when they are bigger.

To raise fish, the oxygen level of your pond becomes a crucial factor. Besides keeping the pond clean and clear of too much animal waste and weeds, the aerator can be essential. Solar pond aerators and fountains are available; they can add essential oxygen for longer periods of time than an emergency aerator (such as a lawn sprinkler) can provide. It is much smarter to prevent a drop in oxygen levels, which can be caused by a heat wave or a nitrogen bloom. The biological controls of the pond are in three levels:

1. Aquatic vegetation, like duckweed. These create oxygen during the day and feed your plant-eating fish.
2. The plant-eating fish eat the plants and create fertilizer and debris. Bass or other predator species may be able to live in a netted-off area to control the fry population.
3. Freshwater prawns live on the bottom, cleaning up debris.

You can eat all of these creatures as well. The prawns need help periodically to clean up the muck, which must be dredged up or it will decrease the oxygen in the

Rules of Successful Aquaculture Management

1. The pH of the water should be 7+.
2. Don't use fish stock with diseases.
3. Choose species you want to eat and you can also sell locally.
4. Make sure to analyze your aquaculture in relation to the elements around it.

water. This precious material should be used as compost or mulch, or used to grow seedlings as historically was done in a chinampa system. Or it can be piped and automated through an aquaponics system.

A pond with lots of fertilizer will grow more algae, and if you want to raise mostly prawns then you might want to leave more of it on the bottom than usual. Typically, a pH of 7+ is necessary to grow fish, and this is regulated by the addition of fertilizer. A semi-fertilized pond grows tilapia. A pond with bass or trout must be cleaner. Fish need to eat 1 percent of their body weight per day. To grow big and fat they must eat 3 percent of their body weight.

Per quarter acre (0.10 ha) of pond surface, it is possible to raise all of this:
- 40 pounds of bass (predator species)
- 80 pounds of catfish
- 120 pounds of bluegill (eats insects)
- 350 pounds of tilapia (eats plankton and plants)

Tilapia is the most versatile and efficient fish to grow, and for this reason it is incredibly popular. The stocking rates per acre are maximum numbers that are used as a guide by fisheries to show how much you can grow if you are only raising a single species in a pond. See the tables that follow for some general polyculture stocking guidelines. These are some common species used in polyculture (please note that regulations vary from region to region, and some exotic species may not be legal in your area):

Tilapia: Tilapia don't like cold temperatures and prefer warm water, but they are fairly hardy and will tolerate most places during the summer. They only take four months to grow from a fingerling to something edible. Duckweed is the best way to feed tilapia. See the chapter on plant species for more information on growing this amazing water plant. They need adequate shade under water lilies and other plants. To harvest, lower a net into the water and spread worms or breadcrumbs on the surface. When they come over to feed, scoop them up with the net. You can winter a few small tilapia in an indoor aquarium or tank. When the weather warms up, release them back into the pond. The stocking rate of tilapia is around three thousand per acre.

Bluegill: It takes about three years for bluegill to grow to an edible size. They prefer warm temperatures and are often used to feed bass. They eat insects, fish eggs, and small crayfish. They also like vegetation for shade and shelter. The stocking rate of bluegill is around five hundred per acre.

Bass: Bass are the most effective method of population control in a pond, because they eat other, smaller fish. They like warm water with lots of plants. The stocking rate is one hundred per acre.

Minnows: Minnows eat mostly algae and some insects, and their principal use is as food for bass and other predators. They should be introduced a year before the bass are brought in, so they can build

up their size and numbers. They can live to be three years old, but are usually eaten before then. The stocking rate of minnows is up to two thousand per acre.

Trout: Trout need cold water, and so they do better in big deep ponds. Alternatively, you can raise them over the winter. The water temperature needs to stay below 60°F (15°C) and above 34°F (1°C). They don't do well with other fish besides minnows, and they eat insects as well so you wouldn't need as many minnows as you would for bass. The stocking rate of trout is four hundred per acre.

Shrimp/prawns: The fewer the shrimp, the larger they grow. Freshwater shrimp need temperatures above 65°F (18°C) but not too hot. They don't survive a cold winter. Shrimp are stocked at a rate of sixteen to twenty-four thousand shrimp per acre depending on how big you want the shrimp to get.

Mussels: Freshwater mussels prefer a temperature of 50–90°F (10–32°C). They eat very tiny planktonic food at the bottom of the pond and act as a natural filter. You need to use local species that like living in still water. Check your regional fish laws for how many and what type you are allowed to collect, and take a few from several ponds or lakes so that you don't impact the population. They can filter up to a gallon of water an hour, so you don't need that many. They can be stocked at a rate of about two hundred per acre as long as they have adequate food to eat.

Catfish: The more catfish eat, the bigger they get. They eat minnows and small fry. Make sure you have at least a thousand minnows per five hundred catfish. Catfish are winter hardy, although over the winter they don't eat or grow much. Catfish don't do well with other species besides minnows, and they prefer a clean pond. The stocking rate of catfish is fifteen hundred per acre.

Crayfish: Crayfish (or crawfish) are another bottom feeder, and they prefer temperatures around 65–85°F (18–30°C). If the temperature drops below 45°F or above 88°F, they will burrow into the ground and go dormant. They are vulnerable to predators, especially when they molt, because they do come up on the shore. They need lots of vegetation, shallow water, and hiding places. The stocking rate of crayfish is two hundred per acre.

Stocking rates depend on two factors: oxygen and food supply. The more you aerate the pond, the more fish you can grow. The more plankton, vegetation, and minnows you have, the more food supply is available. At the same time, the more plants and fish you have growing, the less oxygen will be available and the less you can grow. It is a cycle that you must try to keep in balance. The following tables are illustrations of sample stocking rates in a polyculture system. The rates are intentionally set quite low for better growth rates.

The best rules of thumb used here are: reduce everything at least one-third simply to have enough oxygen; avoid growing species that compete for the same food supply, unless you are experienced.

Stocking Rate Per Acre

Breed	Monoculture
Crayfish	200
Tilapia	3,000
Bluegill	500
Bass	100
Minnows	2,000
Trout	400
Shrimp	16,000
Catfish	1,500

Suggested polyculture scenarios:

Bass Farming

Breed	Stock Rate Per Acre	Stock Rate .25 Acre
Crayfish	60	15
Tilapia	1,000	250
Bass	100	25
Minnows	600	150
Shrimp	5,000	1,250
Mussels	60	20

Catfish Farming

Breed	Stock Rate Per Acre	Stock Rate .25 Acre
Crayfish	60	15
Catfish	500	125
Minnows	1,000	250
Shrimp	5,000	1,250
Mussels	60	20

Aquaponics

It makes sense, once you have become obsessed with raising fish, to turn to aquaponics. Aquaponics is a closed-loop growing system that works a lot like hydroponics, except that instead of chemical fertilizers you are using natural fish fertilizer from live fish directly from your own tanks. This system uses 10 percent of the water that would be used by traditional farming because it is continuously recycled, and it produces more food in less space. This is a relatively new field of production and so there isn't even a lot of information or books out there to guide you.

As a beginner, the best thing you can do is to purchase a predesigned aquaponics system from a reputable company. This way you can skip all the trouble and expense of trial and error that has already been worked out in these systems, and it is well worth it. Once you've bought that small-scale system, you'll have the experience to build your own large-scale system. Greenhouses work best, not only to keep the fish at a constant temperature, but to provide a controlled, well-lit environment for your plants as well. Many people are doing this indoors as well, with some success as long as high-tech LED lights are used.

These are the parts of an aquaponics system:

Grow bed: A strong, waterproof, food-safe container with enough volume for the plant roots as well as a flow area under the plants.

Piping: Pipes that flow into the grow bed have holes for water flow, while other pipes are smaller and push water from or to the fish tanks. These are PVC or CPVC.

Fish tank: A short, wide tank is better for fish because it prevents dead zones and has more surface area. It should also be designed to hold the volume of water it needs to.

Pump: A magnetic drive submersible pond pump with the right flow rate and head pressure.

Sealing: Seals between pipes and tanks include silicone, Uniseals, and bulkhead fittings.

Timers: Automation is essential in the aquaponics process to reduce fatal mistakes. The pump needs to be timed in order to flow the right amount of water.

Grow media: A structure for your plants to sit in that doesn't decompose and is the right size to not clog the pipes. This has to be purchased from a supplier.

Aerator: Something to add oxygen to the water.

BUTCHERING AND PRESERVING MEAT

Putting an Animal Down

There are times when you will have to put an animal down because of injury, age, or illness. In those situations, you may also want to use the resources available in order to not waste anything. While you can't eat the meat, you could make leather or gelatin. Butchering can be sad

▲ This is the reality of owning animals for food. If you can't handle this part, it's probably not a good idea to have them.

and stressful. If you are butchering a wild animal then you may not feel as much remorse for the creature as if you raised it yourself, but it is still sad. If you raised the animal from a baby and it had a name, it will be much more difficult to eat. It doesn't matter if the animal is for milk or eggs or meat, they all deserve love and kindness, and they need to be killed humanely, without terror, with as little pain as possible. Don't kill the animal in its home in front of its family.

It is time to butcher farm animals for meat when the temperature is 40°F during the day, when the pasture doesn't make enough food for the animals, or when a sheep or goat is nine months old or younger. Chickens can be killed anytime.

Decide what you are going to keep of the animal. Hooves can make gelatin, and intestines can be used for catgut (used in stitching up wounds). You can also save the hide for making leather. You will need a butchering knife, a gambrel, a big container to hold guts (like a bucket), a hose connected to running water, and a large clean bowl. It is also a good idea to have a rope or chain to hoist the body, hooked to the gambrel, up high enough for you to work comfortably. Another very good idea is to get an experienced friend to help you and teach you how.

Slaughtering Methods

According to the University of Iowa, there are only a few acceptable methods of killing an animal, and most require a veterinarian:

- Small animals (rabbits/rodents): carbon dioxide, barbiturate overdose, and anesthetic overdose
- Dogs and cats: barbiturate overdose, anesthetic overdose

- Birds: barbiturate overdose, anesthetic overdose
- Farm animals: barbiturate overdose, anesthetic overdose

Notice that it is never recommended to shoot the animal or chop off your bird's head, as most farmers practice. Injecting an animal with drugs before butchering makes it unusable for meat, and it is incredibly expensive. However, before you go out and start shooting or knifing an animal, you must remember that if you get it wrong you will cause a great deal of pain in the last moments of your animal's life. Also, if you stress out a goat before you eat it, the meat won't taste as good. Learn to do it right, or don't do it at all. Practice your aim on an inanimate target first so that you will only have to stab once.

In the country you would shoot a goat with a .22 rifle in the back of the head. A larger animal needs to be shot in the front, hopefully with a hunting rifle. To find the right spot, draw a line from the tip of each ear to the opposite eye, making an X. Shoot at the center of the X. Try very hard to shoot a big animal on the first shot or it will charge toward you.

In the city, you can't shoot a gun, so you have to use a very sharp knife. You will need to quickly and cleanly cut the jugular, the main artery in the neck. First, sharpen your knife to the very sharpest it can be. Take the animal away from its family, far enough that they won't hear it, and tie its legs together. For a goat, the easiest way is to place it on the table with a bucket below and hang the head over the bucket. However, you also need to calm the animal down so that might not be the best position. Once it's calm, grasp the ear firmly and stick the knife in behind the jaw. Pull out quickly, slicing all the veins and arteries.

The Ultimate Guide to Urban Farming

Burying an Animal

Sometimes livestock die from an infectious disease or a disease that prevents you from using it for meat, or if you don't eat your livestock, animals will just get old and die. Eating meat contaminated this way can cause you to get sick or even die. In most places, the requirement is that you dispose of the body on your own property within thirty-six hours after death. The only way to have someone else dispose of it somewhere else is if you hire a licensed rendering company.

Before meat processing got more efficient (or regulated), you could pay a licensed rendering company to get your animal, liquefy it, and turn it into meat and bone meal for feeding to animals. This is not a good choice because of transmission of diseases, and the ethical implications of this option are obvious. Another option is to incinerate the animal in a special high temperature incinerator. These require permits because they stink and use a lot of fuel. Some large farms compost their animals above ground in special facilities, then spread the compost on the fields. For the small homesteader, burial is the best method, but if the ground is frozen you might have to cremate the animal. If you do decide to burn the body, make sure you burn it down to ashes. One good thing about cremation is if the animal was sick, there is no chance of spreading disease.

Here are some rules for burying an animal safely:

- You must dig a hole that will put the animal at least four feet below the surface.
- The land can't be dug up again—it must be kept as an animal graveyard.
- The burial pit must be at least one hundred feet away from anything.

- The soil must be deep and fine textured.
- There should be no danger of groundwater being contaminated.

Immediately after Killing

Note: Pigs need special procedures for cleanliness. Don't follow these instructions for hogs.

1. If the animal is not where you want to butcher it, put a noose around its neck and drag it to where you want. This needs to be done right away because the next step is best done with the heart still beating.

2. Hang the animal upside down from the hind feet and slit the throat (if you haven't already) by sticking your big knife in and pulling it outward. Make sure to sever the arteries and veins. Any time you make a cut, avoid cutting into the hair—instead keep your knife between the flesh and skin and cut out. It is so important to hold the hide away from the meat as you cut (for example, hold it away with your left as you cut with your right). Your hand, the hide and hair should never touch the meat, in order to avoid contamination.

3. If this is an uncastrated male, remove the head and testicles or the meat will be tainted and taste bad. To remove the head, use the slit in the throat to cut all the way around. For a goat, twist the head until the bone snaps. With a larger animal use a meat saw to cut the spine.

4. Make slits between the Achilles tendon and the ankles and insert the gambrel. Remove the front feet, and hoist the animal to a height convenient for work on the animal's rear end.

5. To remove the skin, starting at the slits at the tendons, cut around the foot (cutting out), and be careful not to cut the tendon. Slice a line down each leg from that point in the center of the leg. Then where the two lines meet, slice down the center of the body to the neck.

6. Starting at the junction between the two leg cuts and the body cut, use the skinning knife to separate the skin from the flesh. You will have to pull the skin with one hand as you go so they will come apart.

7. If you are going to save the hide, be careful as you go. Skin the whole belly, then work around the legs from front to back.

8. Start the top of the Y (the junction of the cuts you made earlier) and skin up over the crotch. This is the tightest spot so be very careful if you are saving the hide. If you leave the fat on the body it will be easier to skin.

9. Skin over the anus to the tailbone. Pull the tail sharply and it will separate from the spine. The rest of the animal will be easier to skin. Raise or lower it to be a comfortable height.

10. To skin the forelegs, start on the outside of the leg and work around to the front. Then skin the neck and inner forelegs and the skin should come off.

11. Cut around the anus with the sharp pointed knife, being careful not to poke any holes in the intestine. When the anus is free, pull it out slightly and tie it off, unless it is a goat (not necessary for a goat).

12. Cut down the belly (from the inside out), holding the guts away from the point of the knife with your other hand. Cut through the belly fat down to the sternum, and then cut the meat between the legs.

13. Cut out the penis if it has one, then place a very large container underneath to catch the guts, which will be bulging out of the hole you just made. If this is a ruminant animal, green liquid may flow from the neck, but this is just cud. Don't let any cud get into your container.

14. Cut through the fat surrounding the guts and sever any tissue connecting them to the rear wall of the body cavity. Pull the anus through from the inside, and then take it out of the body cavity through the big hole. Take your time with this step and go slow—if you poke a hole in any of the intestines or the bladder, it will contaminate the meat because liquid is still in there.

15. Pull the intestines and bladder out of the body. Just reach in and lift them over the sternum and out (they go into the big container), and most of everything will be hanging out of the body. Strip off as much belly fat as you can, which can be fed to chickens. Cut out the organs you want to keep and put them into another bowl that you have handy (the liver, kidneys, etc.).

16. Cut the flesh that is left, which is connecting the stomachs to the body, and let them fall into the big bucket. Cut out the diaphragm and the connecting tissue behind the lungs and heart to remove them. Separate the heart from the lungs, and squeeze it to get out the blood. Put the heart in the bowl to keep. Put the lungs in the bucket.

17. At the neck, cut out the windpipe and make sure the hole is clear all the way through the body cavity. Hose the whole thing down with cold water.

The Ultimate Guide to Urban Farming

18. To get the tongue, cut under the jaw in the soft space in the middle. Then reach in and cut the tongue loose at the base. To get the brain, saw it in half with your meat saw. If you want to save the head for head cheese, skin it, remove ears, eyes, nose, and anything else that is not meat or bone, and brush the teeth to clean them.

Cutting Up a Goat

1. Cut behind the shoulder blade to remove the front legs, then cut off the bottom half of the leg at the elbow. The bottom half is not good for anything but soup bones. The shoulder can be packaged as it is, or the meat can be taken off the bone and rolled and tied for roast, or it can be chopped for stew meat or stir fry.
2. Take as much meat from the neck as you can and use it for soup. Saw through the backbone between every rib to get chops, or take the bone out completely, or just cut out the whole muscle bundle along the backbone. When the whole muscle is taken it is called backscrap and is the best-tasting meat.
3. Use the meat saw to cut the ribs from the backbone, then cut the ribs in half with your knife and package. Cut out the meat under the backbone—this is the tenderloin.
4. Cut off the rear feet, then cut off the legs at the knee. The bottom half of the leg is the shank. Remove the leg at the pelvis and use it for roast. Take out the bone now or package as it is.
5. Go over all the bones and get any last bits to be put into sausage and jerky.

How to Kill a Chicken

1. Catch the chicken, either by sneaking up on it at night or by catching it some other way. Have a large bucket ready for catching blood, and if you are going to remove feathers, have a large pot or bucket starting to boil.

▲ This is the Whizbang chicken plucker designed by Herrick Kimball.

2. There are many ways to kill a chicken. A common way is to do it with an ax on a chopping block. Hold the chicken with one hand and use a heavy-headed ax to chop the head off in one blow. The chicken may move its head at the wrong time, but the goal is to cut as close to the head as possible. Another way is to hold the chicken just below the head and swing it so the body twirls around, and on the third swing the head will separate from the neck. Or to prevent flopping, tie the chicken by the legs upside-down to a tree branch. Cut off the head with a sharp knife. The least messy way is to cut off the top of a milk jug, two inches below the handle. Nail it to a wall, with the small end down. Pick up the bird by the ankles. Put it head downward into the jug so that its head pokes out the small hole (that used to be a pouring spout), hold the head and stretch out the neck a bit and cut at the base of the head with a sharp knife. A .22 will kill a chicken also.

3. If you used a method in which the bird can't move, the bucket should be ready under the chicken to catch the blood. If you used the chopping method, you will need to quickly stick the bird into the bucket, or the chicken will flop around all over for a long time, scattering blood everywhere. Let the blood completely drain out.

4. If you just skin a chicken you do not have to go through all the work needed to remove feathers. To remove feathers, while the bird is draining, make sure the boiling water temperature is between 130°F and 180°F (hotter is better). Lay out newspapers or other cover to pluck on.

5. Grasp the bird by the ankles and immerse the body in the hot water. If you are using a roasting-style pan, soak the breast first and flip it. Every feathered part should have an equal time in the water, including the feathered knees. It should take about thirty seconds.

6. Put the bird on the newspaper (but don't ever let the skin touch the lead-filled ink), and start picking out feathers as soon as you pull it out. Start with the wing and tail, then the body and then the pinfeathers. The big ones need to be pulled in the direction they grow.

7. You will be left with a bird covered in down and fine hairs. Use a gas stove flame, candle, propane torch, or alcohol burner to singe these off. Make sure you can't set anything else on fire.

8. Now the chicken is cleaned or drawn. Put it in very cold water to cool it, and either get ready near a faucet or get a bowl of cold water. Also get a sharp knife, a cutting board, a box lined with wax paper or a bucket, and a container to put chicken pieces in.

9. When looking at the bird, if it feels very light and the muscles are thin, and the innards look strange and have abscesses, then there is a strong possibility the bird had tuberculosis (or TB), so don't eat it!

10. Cut off the head and throw it out. If there is no head, cut off the top of the neck if it got dirty, and discard.

11. Feel for the knee joint where the scaly leg joins to the feathered knee, bend the knee and cut across the joint until the foot is off.

12. Put the chicken on its back with the rear end facing you. Cut across the

abdomen from one thigh to the other, being very careful to not cut into any intestines. Reach inside between the intestines and the breastbone until you reach the heart, and gently loosen the membranes that hold the innards to the body wall. The gizzard, heart and liver will come out in one big mass; put them in the box or bucket. Watch out for the green sac embedded in the liver; if it breaks it will make the chicken taste bad.

13. Scoop out the lungs, which are stuck to the inside of the ribcage, with your fingers and put them into the box or bucket. Cut around the vent (the anus) with lots of room so you don't cut into the intestine. Throw that into the box or bucket.

14. The kidneys will be stuck inside the cavity against the back. With your fingers, scoop out as much soft kidney tissue as you can, but you won't be able to get all of it, and that's OK.

15. Cut the skin down the whole length of the back of the bird's neck. The crop (the thin skinned pouch on the bird's esophagus) tears really easily, so you will have to pull the skin away. Pull the crop, esophagus, and windpipe out of the bird and put it in the box or bucket.

16. Dispose of the innards by burning, composting, or if they had no chicken diseases, feeding them back to the chickens. Don't feed to dogs or cats or they will start killing chickens. If you want to save the giblets (heart, liver, and gizzard), cut the gallbladder off of the liver, rinse the liver, and put it in a giblet bowl. Cut the gizzard open along its narrow edge three-quarters of the way around, and dump out food and grit. Peel off the yellow inner lining

in the gizzard by separating it at the edge you cut at and pull it out. Rinse the gizzard and put it in the bowl, then cut the heart away from the arteries, rinse and save it. Cut the neck off at the base and put it in the bowl also. Wash the feathers and save.

17. You can clean the outside of the chicken with soap and water if you rinse well, which removes any remaining dirt.

How to Kill Any Other Bird

Turkeys can be killed the same as chickens, except the bird is five times bigger so it will take longer. To scald a turkey, it is easy to use a metal garbage can and heat the water to 140–180°F for thirty to sixty seconds. To remove turkey pinfeathers, put the bird under a faucet and use pressure and rubbing to remove them. Ducks are difficult to catch because they are easily injured, so grasp the bird by the neck and pull its body into your armpit, holding the wings down with your arm. Keep the duck away from your face, and kill just like a chicken. Hold geese the same way as a duck, but make sure that you are holding the neck behind the head so you can keep the beak away from you. To scald a goose, heat the water to 155°F, and soak for as long as four minutes. To clean water fowl, you will also need to remove the oil sack on the tip of its rear end and all the yellow area around it.

Killing a Rabbit

1. Hold the rabbit's hind legs as high as your chest with your left hand, and hold it around the neck with the right. Pull the head down and while bending the head backwards as hard as you can until you feel the neck snap. The other

method is to whack it in the back of the head as hard as you can.

2. Hang it by the hind feet with the feet spread as wide apart as the body.
3. Cut the skin off from the hock joint of each hind leg and peel back so you can see the Achilles tendons, and stick hooks, ropes, chains, or a gambrel to hook in the tendon.
4. Cut off the front feet and tail. Use a small sharp knife to cut the skin between the legs, starting with the hock joint on the inside of the leg and working up the inside of the other leg to the other hock joint.
5. Cut the skin from each front leg to the neck, and peel the skin off the body starting at the hind hocks. It should peel like a banana, so that it will be inside out. Cut around the anus and any place that the skin does not come off easily.
6. Put the skin aside for tanning. Cut around the anal vent and down to the breastbone down the middle of the belly. Don't cut into any intestines!
7. Pull out the intestines, reach into the chest and pull out the heart and lungs. Look at the liver, and if it has white spots or any discoloration, it has in infection and you can't eat it and

---

Meat grinder plate sizes:

⅛ inch: hamburger, bologna, franks

¼ inch: hamburger, salami, pepperoni

3/16 inch: course-ground hamburger, breakfast sausage

⅜ inch: chili meat, first sausage grind

½ inch: chili meat, vegetables

---

neither can any animal. If it has white cysts on its stomach or intestines, it had tapeworm and it must be cooked very well to be eaten.

8. If you want to save the liver, cut away the gall bladder very carefully, and don't spill the gall bladder. Wash the liver well, and wash the rest of the body.
9. Take the rabbit from hanging up, then cut off both front legs at the shoulder and both hind legs at the hip.
10. Cut through the ribs on both sides parallel to the backbone. Divide the rest of the back into two or three pieces. Cut away the meat from the hind legs and haunches or loins.

Aging Meat

A goat should age one week at 40°F, and longer if it is colder. A larger animal needs to be hung at least two weeks at 40°F. A goat can be hung whole, but a big animal needs to be cut in half. To halve it, have someone help you. Face the belly while your helper holds the body and helps guide the saw from the back.

How to Fillet a Fish

Note: This technique can be used for a medium-sized fish such as coral trout, barramundi, or salmon and can be used on any fish with scales.

1. Put the fish on a very clean, flat surface belly down. If the head is not removed already, hold onto its head with the hand not holding the knife (if you are right-handed, then the head would be toward your left). You will need a thin flexible knife and a broad flat knife, both sharp.
2. Use the thin knife to cut downward just behind the head in a diagonal line until

you reach the backbone, and run the knife along the top of the bone until you reach the top fin nearest the tail. Use a sawing motion and do not try to cut off too much fillet at this point.

3. When you're near the tail, hold the knife flat against the backbone and push the point through the side of the fillet. With the knife sticking out the other side, cut through the remaining fillet toward the tail. Peel the fillet back with one hand while cutting with the other using small slicing motions. Turn the fish on its side with the backbone facing you.

4. Starting just below the Y bones coming off the backbone, cut one-quarter inch deep the length of the fish until you reach the point you stopped with the first fillet.

5. Cut deeper this time, skimming along the edge of the ribs from front to back again, stopping at the same point as before.

6. Make a cut along the top of the belly by following the white line in the skin and lift off the fillet as you cut.

7. Turn the fish over and repeat steps 3–7 on the other side. After you're finished, there will be sections of meat behind the rear dorsal fins. This portion does not have Y bones in it, so run your knife along the backbone all the way to the tail to remove a boneless fillet.

8. Repeat on the other side and you should have a total of five boneless steaks.

9. To remove the skin (this is optional): Hold the skin in one hand and use the flat knife to slice a small portion of the flesh away from the skin. Cut a finger hole in the skin and hold it by the hole.

10. Use the knife to gently remove the skin by holding the knife at an angle and pulling with the other hand. Do not cut or push with the knife.

The teeth in the fish can be sharp so be careful not to stick your hand in there or you can get cut. Don't poke yourself with the spines in the fins. Freeze or can any fish that you are not going to use right away.

Rendering Tallow

1. Set up a big pot over a heat source. It is best to do this outside because of fire risk and the greasy coating this process creates. Build a fire without high flames or flying embers under the pot, or the tallow may burst into flame.

2. Have a lid handy that fits completely over the pot so if the tallow does catch fire, simply cover it to seal out oxygen. Do not try to put it out with water.

3. Fill a big pot ¾ full of chopped fat (goat, deer, elk, or whatever animal). The chunks should be no bigger than ½–1 inch in size). Cover the fat with water and bring to a rolling boil.

4. As the water boils away the tallow will begin to be extracted from the fat. Keep up the boil the whole time, and when most of the water is gone you will see the mixture change.

5. Watch the pot closely. You will see the bubbles becoming smaller and less violent, and the color will change from light, muddy brown to a dark, clear liquid.

6. The fat is rendered when it turns into brown crisps that look like bacon, and it will smell like bacon. This will take about three hours, and when the water is completely evaporated a light, white smoke will come off it. Take the pot off the flame immediately or it can catch fire, and be very careful not to spill or it can cause an explosion.

Neat's-Foot Oil

Boil the hooves, horns, and scrap pieces of hide for a few hours. Let it cool and the oil will rise to the top. Use as it is or mix with fat.

---

7. Let cool 10 minutes then strain out the impurities. When it is still warm you can use it to make tallow candles in a candle mold.

Gelatin

1. Use goat or sheep leg bones, or chicken head, feet, necks or backs (3 pairs of feet make 1 pint of gelatin). Put them in a pot and cover with water.

2. Simmer the bones for at least 4 hours. Goat and sheep should be boiled for 6–7 hours, while chickens need a bit less.

3. Let the bones cool, then skim off the fat from the top of the water (see Neat's-Foot Oil, above), and the sediment from the bottom, and remove the bones. The liquid left over is the gelatin.

4. To clarify the gelatin, heat until the gelatin melts. For every quart add ½ cup sugar or honey, and the shells and slightly beaten egg whites of 5 eggs. Stir it in, then stop stirring when it gets any hotter.

5. Let it boil 10 minutes, add ½ cup cold water, and boil 5 minutes more. Remove from heat, cover with a lid and let it sit in a warm place for 30 minutes.

6. Get a large clean dishcloth and fold it to make several layers. Soak it in hot water and ring it out, then strain the gelatin through the cloth. You will have to squeeze it through the cloth (hold the cloth like a bag).

7. If it doesn't get clear, strain it through again. To set up, or harden the gelatin after using it in a recipe, put it in the fridge. If it doesn't set up that means you added too much liquid. Pineapple, papaya, mango, and figs all need to be cooked before using or they will also prevent the gelatin from setting up.

Salting Meat

You could smoke meat instead, which tastes delicious, but smoking takes a smokehouse and several weeks of time and constant vigilance. Salting is a practical and efficient method of preserving meat with very little effort.

1. Clean the meat, and cut off anything you don't like. Save the fat to make clarified fat. Dry the meat with a clean cloth, and cut it into smaller strips so that it will be easier to make sure that the meat in the middle is preserved. Rub spices into the strips, and then rub tons of salt into them until you can't rub any more. There are salt curing products available made just for this purpose, but these contain sodium nitrite, which you probably don't want to use.

2. When you've rubbed in as much salt as you can, cover it in a layer of salt to coat it. Hang it up in a place that stays consistently 59°F (15°C) for at least three weeks, checking often for spoilage. A basement or cold storage is ideal. It should stay edible for at least a few months. The way this works is that salt dissolves into the water in the meat, preventing bacteria from growing if the balance is greater than 3.5 percent salt to water. Ideally, it should be over 10 percent salt. You can't really control the percentage but if you just rub so much salt into the meat that it just won't accept any more, you can be fairly sure that you've got it.

3. When you are ready to cook it, just wash off the salt well. You might have to soak it a little bit to get it all out.

Clarified Fats and Butters

Fat is useful and healthy, as long as it's used the right way. There is a big difference between the fats found in factory foods and the natural fats found in homegrown meat or dairy. You will still want to remove excess fat when you are cooking, but it serves a useful purpose.

Lard is an unpleasant word that is synonymous with clarified fat. It is fat that is cut up, liquefied and filtered. When it cools, it becomes a block of lard. Often several types of fats are mixed together, and you can even add a small amount of vegetable oil. To make clarified fat, save all the fats from chopping up meat and store them in the freezer for when you are ready to process them. When you are ready, put them into a saucepan and simmer on low for a few hours until the fat is liquid. If you would like to make your own bullion cubes out of it, cut up onions, carrots, leeks, turnips, herbs, and spices, and add salt and pepper to the fat. Once it is liquefied, pour it through cheesecloth and allow it to cool. It should last years, and can be used in soups and stews, or for greasing a frying pan.

Clarified butter, or *ghee*, lasts much longer than regular butter. Once made into ghee, it can sit at room temperature for months without going bad. The butter is melted at a very low temperature until it is completely melted. Don't stir it, but you can raise the heat slightly so that it steams a little. Don't let it turn brown. The solids will rise to the surface, which will need to be skimmed off. Eventually (usually hours later), it will be a golden color and completely clear. Pour it into a container and when it is solid and cool, put the lid on tightly. Clarified butter is used in cream sauces and for frying.

Egg Storage

If you have quite a few chickens, they will probably produce more eggs than you can eat. Eggs can actually last a long time on their own. In the fridge in a regular egg carton they last about six weeks, and in a plastic bag can last two months. Cold storage, pickling, drying, and freezing can extend that time and possibly get them through the winter when the chickens aren't laying.

For cold storage, pack freshly gathered eggs into a wood, plastic, or ceramic container in sawdust or oatmeal with the small end down. Don't store the eggs near anything smelly like onions. They must be stored at 30°F to 40°F (–1°C to 4°C) in fairly high humidity, and will last about three months.

To pickle eggs, first hard-boil them, cool immediately in cold water, and remove the shells. Put them into wide-mouthed jars. Soak them in a brine of ½ cup of salt per 2 cups of water for 2 days. Pour off the brine. In a saucepan heat:

- 1 quart vinegar
- ¼ cup pickling spice
- 2 cloves garlic
- 1 tablespoon sugar

Bring this mixture to boil and pour it over the eggs. Screw the lid on tightly and leave them alone for seven days to cure before eating. Pickled eggs will last four to six months in the fridge or a properly chilled cold storage.

The easiest method of storing eggs is freezing. Use only fresh, clean eggs that you didn't have to clean yourself. This means eggs that happened to not get any dirt or manure on them. Crack the eggs and put their contents into the freezer container bag. Only freeze as many eggs per bag as you will use at one time

because you can't refreeze the eggs once you thaw them. Stir the eggs together without whipping in any air, and add

1 tablespoon of sugar OR

½ teaspoon of salt per cup of egg

They will keep for eight months in the freezer.

Drying is a somewhat more time-consuming method that extends the life of your eggs. Compared to other methods it may not be worth it since they will only last three to four months, which is as long as simply putting them into cold storage. Nonetheless, you might still want to do it. Crack very fresh eggs and beat them well in a bowl. Pour them into a drying surface that is lined with plastic or foil, no more than an eighth of an inch thick. Plates work for solar drying or you can use a dehydrator. In an oven or dryer, dry at 120°F (49°C) for 24–36 hours, then turn the egg over, remove the plastic or foil, break it up and dry for 12–24 more hours. In the sun, it will take five days until they are dry enough to break easily when touched. Grind the egg into a powder and use in baking, or reconstitute by adding an equal amount of water (half cup of egg powder to a half cup of water).

Lard is the most effective method, if you have lots of lard. Use very fresh, clean eggs and dip in melted lard. Lay them out to dry, then buff them gently with a clean towel to remove any excess and to ensure that the lard is spread all over the egg. Then pack the eggs in salt in a large bucket so that no eggs touch each other. Put the bucket in a cool place, and stored this way they will last six months to a year.

Bibliography

BlueScope Water. (n.d.) Rain Water Tanks – Life Cycle Analysis. *BlueScope Steel Australia*. Retrieved from http://www.bluescopesteel.com.au/building-products/rainwater-harvesting/life-cycle-analysis-for-rainwater-tanks

Bohnsack, Ute. (n.d.) Companion Planting Guide. *South East Essex Organic Gardeners*. Retrieved from http://www.gb0063551.pwp.blueyonder.co.uk/seeog/companion/

British Columbia Ministry of Public Safety and Solicitor General. (n.d.) Flood Proofing Your Home: Minimize Damage if a Flood Strikes Your Family Home. *Emergency Management BC*. Retrieved from http://www.pep.bc.ca/hazard_preparedness/flood_tips/Floodproof.pdf

Burns, Russel and Barbara Honkala. (1990) Black Walnut. *Silvics of North America*. US Department of Agriculture, Forest Service. Retrieved from http://www.na.fs.fed.us/pubs/silvics_manual/volume_2/juglans/nigra.htm

Carpenter, Novella and Rosenthal, Willow. (2011) *The Essential Urban Farmer*. Penguin, New York.

Clarke, S. (2003) Electricity Generation Using Small Wind Turbines at Your Home or Farm. *Engineering Factsheet*. Ontario Ministry of Agriculture, Food and Rural Affairs. Retrieved from http://www.omafra.gov.on.ca/english/engineer/facts/03-047.htm

Clouse, Carey (2014) *Farming Cuba: urban farming from the ground up*. Princeton Architectural Press, New York, NY.

Coleman-Jensen, Alisha, M. Nord, M. Andrews, S. Carlson (Sep. 2011) Household Food Security in the United States 2010. USDA. http://www.ers.usda.gov/publications/err-economic-research-report/err125.aspx

Cook, Howard. (2011) *The Principles and Practice of Effective Seismic Retrofitting*. Bay Area Retrofit. Retrieved from http://www.bayarearetrofit.com/PDFs/design_book.pdf

Cunningham, Sally Jean. (1998) *Great Garden Companions: A Companion Planting System for a Beautiful, Chemical-Free Vegetable Garden*. Rodale.

Denzer, Kiko. (2001) Build Your Own Wood-fired Earth Oven. *Mother Earth News*. Retrieved from http://www.motherearthnews.com/Do-It-Yourself/2002-10-01/Build-Your-Own-Wood-Fired-Earth-Oven.aspx?page=5

Deveau, Jean Louis. (2002) Raising Quail for Food in Fredericton, New Brunswick, Canada. *Urban Agriculture Notes*. Retrieved from http://www.cityfarmer.org/quail2.html

Down Garden Services. (2011) Companion Planting. *Down Garden Services*. Retrieved from http://www.dgsgardening.btinternet.co.uk/companion.htm

EPA. (n.d.) Sources of Greenhouse Gas Emissions. EPA. Retrieved from http://www3.epa.gov/climatechange/ghgemissions/sources/agriculture.html

Exploratorium. (n.d.) Basic Sourdough Starter. *Science of Cooking*. Retrieved from http://www.exploratorium.edu/cooking/bread/recipe-sourdough.html

Faires, Nicole. (2013) *Food Tyrants: Fight For Your Right to Healthy Food in a Toxic World*. Skyhorse. New York, NY.

Faires, Nicole. (2011) *The Ultimate Guide to Homesteading: An Encyclopedia of Independent Living*. Skyhorse Publishing.

FAO. (Jun. 2006) Country Profile: Food Security Indicators Cuba. Food and Agriculture Organization of the UN. http://www.fao.org/fileadmin/templates/ess/documents/food_security_statistics/country_profiles/eng/Cuba_E.pdf [Aug. 4, 2012]

Forcefield. (2008) The Bottom Line About Wind Turbines. *Otherpower.com.* Retrieved from http://www.otherpower.com/bottom_line.shtml

Fowler, D. Brian. (2002) *Winter Wheat Production Manual: Chapter 10 – Growth Stages of Wheat.* University of Saskatchewan College of Agriculture and Bioresources. Retrieved from http://www.usask.ca/agriculture/plantsci/winter_cereals/Winter_wheat/CHAPT10/cvchpt10.php

Fraas, Wyatt. (n.d.) Beginning Farmer and Rancher Opportunities. *Center for Rural Affairs.* Retrieved from http://www.cfra.org/resources/beginning_farmer

Fukuoka, Masanobu. (1985) *The Natural Way of Farming: The Theory and Practice of Green Philosophy.* Japan Publications.

Garden Action. (2010) Plum Tree Care. *GardenAction.* Retrieved from http://www.gardenaction.co.uk/fruit_veg_diary/fruit_veg_mini_project_march_2_plum.asp#plum_start

Gasparotto, Suzanne. (n.d.) Organically – Raised Goats. *Onion Creek Ranch.* Retrieved from http://www.tennesseemeatgoats.com/articles2/organicGoats06.html

Golden Harvest Organics. (2011) Companion Planting. *Golden Harvest.* Retrieved from http://www.ghorganics.com/page2.html

The Greenhorns. *Guidebook for Urban Farmers (Sep. 2010).*

The Greenhorns. http://www.thegreenhorns.net/wp-content/files_mf/1335219697greenhorns_guide_sept2010_web.pdf [Apr. 30, 2015]

Hackelman, Michael and Claire Anderson. (2002) Harvest the Wind. *Mother Earth News.* http://www.motherearthnews.com/Renewable-Energy/2002-06-01/Harvest-the-Wind.aspx

Hart, Kelly. (n.d.) Frequently Asked Questions. *Earthbag Building.* Retrieved from http://www.earthbagbuilding.com/faqs.htm

Hemenway, Toby. (2009) *Gaia's Garden: A Guide to Home-Scale Permaculture, Second Edition.* Chelsea Green.

Hill, Stuart B. (1975) Companion Plants. *Ecological Agriculture Projects.* Retrieved from http://eap.mcgill.ca/publications/EAP55.htm

Holmgren, David. (2009) *Permaculture: Principles & Pathways Beyond Sustainability.* Holmgren Design Services.

Hooker, Will. "Introduction to Permaculture." North Carolina State University, Distance Education. 18, Mar. 2011.

Jacke, Dave & Eric Toensmeier. (2005) *Edible Forest Gardens Volume 1: Vision & Theory.* Chelsea Green.

Jacke, Dave & Eric Toensmeier. (2005) *Edible Forest Gardens Volume 2: Ecological Design and Practice for Temperate Climate Permaculture.* Chelsea Green.

Jason, Dan. (2011) Growing Amaranth and Quinoa. *Salt Spring Seeds.* Retrieved from http://www.saltspringseeds.com/scoop/powerfood.htm

Jenkins, Joseph. (2005) *The Humanure Handbook: A Guide to Composting Human Manure.* Joseph Jenkins.

Kitsteiner, John. (2011) Permaculture Projects: Coppicing. *Temperate Climate Permaculture.* Retrieved from http://www.tcpermaculture.com/2011/06/permaculture-projects-coppicing.html

Learmonth, Pat. (n.d.) *Accessing Land for Farming in Ontario: A guidebook for farm seekers and farmland owners.* http://www.farmstart.ca/wp-content/uploads/Accessing-Land-for-Farming-in-ON-Guidebook-REV4.pdf [May 31, 2015]

Luzadis, V.A. and E.R. Gossett. (1996) Sugar Maple. Pages 157–166. *Forest Trees of the Northeast.* Cornell Media Services.

Lynch, William. (2002) Fish Species Selection for Pond Stocking. *Ohio State University Extension Factsheet.* Retrieved from http://ohioline.osu.edu/a-fact/pdf/0010.pdf

Madison, Deborah. (1999) *Preserving Food Without Freezing or Canning: Traditional Techniques Using Salt, Oil, Sugar, Alcohol, Vinegar, Drying, Cold Storage and Lactic Fermentation.* Chelsea Green.

Mars, Ross. (2005) *The Basics of Permaculture Design.* Chelsea Green.

Midwest Permaculture. (2011) About the PDC Certificate. *Midwest Permaculture.* Retrieved from http://midwestpermaculture.com/about/certification/

Missouri Department of Conservation. (1999) Fathead Minnows in New Ponds and Lakes. *Aquaguide.* Retrieved from http://mdc.mo.gov/sites/default/files/resources/2010/05/4890_2843.pdf

Mollison, Bill. (1988) *Permaculture.* Tagari.

Mollison, Bill and Reny Slay. (2009) *Introduction to Permaculture.* Tagari.

Mohamed Lahlou. (2000) Slow Sand Filtration. *Tech Briefs.* Retrieved from http://www.nesc.wvu.edu/pdf/DW/publications/ontap/tech_brief/TB15_SlowSand.pdf

National Good Food Network. *Food Hub Center.* Retrieved from http://www.ngfn.org/resources/food-hubs

New Brunswick Department of Environment. (2011) Composting. *Backyard Magic: the Composting Handbook.* Retrieved from http://www.gnb.ca/0009/0372/0003/index-e.asp

North Dakota State University. (2011) Insulation. *Bioenvironmental Engineering.* Retrieved from http://www.ageng.ndsu.nodak.edu/envr/Insulatn.htm

Nova Scotia Museum. (n.d.) Lupin or Lupine (Lupinus Species). *The Poison Plant Patch.* Retrieved from http://museum.gov.ns.ca/poison/?section=species&id=99

OK Solar. (2011) Angle of orientation for solar panels & photovoltaic modules. *OkSolar.com Technical Information.* Retrieved from http://www.oksolar.com/technical/angle_orientation.html

Olsen, Ken. (2001) *Solar Hot Water: A Primer.* Arizona Solar Center. Retrieved from http://www.azsolarcenter.org/tech-science/solar-for-consumers/solar-hot-water/solar-hot-water-a-primer.html

Patterson, John. (n.d.) Solar Hot Water Basics. *Homepower Magazine.* Retrieved from http://homepower.com/basics/hotwater/

Pennsylvania Fish & Boat Commission. (n.d.) Using mussels to "clean" a pond. *Q & A.* Retrieved from http://www.fish.state.pa.us/images/pages/qa/misc/mussels_pond.htm

Permaculture Institute. (2008) *Permaculture Design Certificate Course Outline.* Retrieved from http://www.permaculture.org/nm/images/uploads/PDC_cert_book_.pdf

Permaship. (2011) Graywater in the Garden. *Permaculture Projects.* Retrieved from https://sites.google.com/site/permaship1/permaculture-practice/graywater-garden

Permaship. (2011) Permaculture Pond. *Permaculture Projects.* Retrieved from http://sites.google.com/site/permaship1/permaculture-practice/permaculture-pond

Pushard, Doug. (2010) Rainwater Harvesting – Pumps or Pressure Tanks. *HarvestH20.* Retrieved from http://www.harvesth2o.com/pumps_or_tanks.shtml

Rodale Press. (1977) *The Rodale Herb Book: How to Use, Grow, and Buy Nature's Miracle Plants.* Rodale.

Rolex Awards. (2005) Ancient technology preserves food. *Rolex Awards.* Retrieved from http://www.rolexawards.com/en/the-laureates/mohammedbahabba-the-project.jsp

Rombauer, Irma and Marion Becker. (1975) *The Joy of Cooking.* Bobbs-Merrill.

Roose, Debbie. (2011) Selling, Eggs, Meat, and Poultry in North Carolina: What Farmers Need to Know. *Growing Small Farms.* Retrieved from: http://www.ces.ncsu.edu/chatham/ag/SustAg/meatandeggs.html

Royal Horticultural Society. (2011) Plums: pruning. *Gardening for all.* Retrieved from http://apps.rhs.org.uk/advicesearch/Profile.aspx?pid=339

Schafer, William. (2009) Making Jelly. *University of Minnesota Extension.* Retrieved from http://www.extension.umn.edu/distribution/nutrition/dj0686.html

Solar Cooking International. (2011) Solar Cooking Info. *Solar Cookers World Network.* Retrieved from http://solarcooking.wikia.com/wiki/Solar_Cookers_World_Network_(Home)

Spiers, Adrian. (1997) Pruning to prevent silverleaf. *The Horticultural and Food Research of New Zealand.* Retrieved from http://www.hortnet.co.nz/publications/hortfacts/hf205016.htm

Stark, Kevin. (n.d.) "Anthropology 387: The Aztecs." *Pacific Lutheran University, Anthropology.* Retrieved from http://www.plu.edu/~starkkl/anthropology-387/home.html

University of Vermont, Extension. Direct Market Price Report: Sept 1–Sept 19, 2013. Retrieved from: http://www.uvm.edu/farmpricing/sites/default/files/report-pdfs/price_report_9-22-2013.pdf

USDA, NRCS. (2011) The PLANTS Database. *National Plant Data Team.* Retrieved from http://plants.usda.gov

USDE. (2011) Microhydropower Turbines, Pumps and Waterwheels. *Energy Basics.* Retrieved from http://www.eere.energy.gov/basics/renewable_energy/turbines_pumps_waterwheels.html

USDE. (2011) Natural Fiber Insulation Materials. *Energy Savers.* Retrieved from http://www.energysavers.gov/your_home/insulation_airsealing/index.cfm/mytopic=11560

USGS. (2011) The Water Cycle. *USGS Water Science for Schools.* Retrieved from http://ga.water.usgs.gov/edu/watercyclesummary.html

Virginia Water Resources Research Center. (2011) Vegetated Emergency Spillway. *Virginia*

Stormwater BMP Clearinghouse. Retrieved from http://vwrrc.vt.edu/swc/NonPBMPSpecsMarch11/Introduction_App%20C_Vegetated%20Emergency%20Spillways_SCraftonRev_03012011.pdf

Weinmann, Todd. (n.d.) Companion Planting. *Cass County Extension, North Dakota State University.* Retrieved from http://www.ag.ndsu.edu/hort/info/vegetables/companion.htm

Wheaton, Paul. (2011) Permaculture Articles. *Rich Soil.* Retrieved from http://www.richsoil.com/

White, Mel. (1976) Cure Your Own Olives. *Mother Earth News.* Retrieved from http://www.motherearthnews.com/Real-Food/1976-01-01/Cure-Your-Own-Olives.aspx

Yocum, David. (n.d.) Design Manual: Greywater Biofiltration Constructed Wetland System. *Bren School of Environmental Science and Management, University of California.* Retrieved from http://fiesta.bren.ucsb.edu/~chiapas2/Water%20Management_files/Greywater%20Wetlands-1.pdf

Index

Acacia, 162, 224, 242

Acidic soil, 69, 72

Agricultural hub, 9

Alder, 162

Alfalfa, 82, 162, 168, 225

Alkali soil, 69, 70

Almond, 162, 168–169

Amaranth, 90, 162, 169, 225

American yellowwood, 162

Animal

    Behavior, 219–220

    Biosecurity, 220

    Burial, 267

    Butchering, 265–266

    Health, 220

    Housing, 42–43

    In garden, 38

    Planning, 219

Annuals, 104–105

Apple, 90, 137, 138, 145, 169–170

Apricot, 170

Aquaculture, 39, 260–265

Aquaponics, 39, 264–265

Arid zone, 44

Arthropods, 67

Artichoke, 132, 167, 190

Arugula, 32, 167, 170

Asparagus, 71, 90, 132, 162, 171

Autumn olive, 162, 171, 225

Avocado, 171

Azolla. *See* Duckweed

Bacteria, 67, 136

Bamboo, 132, 162, 172

Basil, 32, 90, 141, 167, 172

Bean, 32, 88, 90, 128, 162, 167, 173

Bean sprout, 132

Bearberry, 162, 173

Bee balm, 141, 174

Bees

    Artificial pollen, 244, 245

    Equipment, 243

    Handling, 244

    Health, 245

    Honey, 247

Beet, 32, 71, 91, 128, 167, 174

Beet greens, 132

Berries, 28, 153–155

Biomass, 58

Biosecurity, 220

Bi-rotation, 26

Black-eyed peas, 132

Black locust, 162

Black tupelo, 162

Blackberry, 175

Blanching, 131, 132

Blueberry, 175

Borage, 91, 175

Boxed bed, 73, 74

Braiding. *See* Dry bed

Broad bean, 104, 132

Broccoli, 32, 91, 132, 167, 175–176

Brussels sprout, 32, 132, 146, 167, 176

Buckwheat, 160, 176–177, 225

Bunchberry, 162, 177

Business

    Cooperative, 13, 14

    Crop value, 16, 26, 32–33

    CSA, 16–17, 28–30

    Funding, 33–34

    Group decisions, 12–13

    Investment, 14

    Plan, 25, 31–32

    Profitability, 14

    Record keeping, 17–19

    Structure, 13–17

Butchering, 265

Butter, 254, 275

Butter bean, 132

Cabbage, 32, 91, 105, 132, 145, 146, 167, 177, 178

Canning, 142–144

Carbon/nitrogen ratio, 80–81

Carob, 162, 177–178

Carrot, 32, 90, 92, 105, 128, 132, 145, 146, 167, 178

Cash crops, 27, 28, 153

Cattail, 162, 179

Cauliflower, 32, 132, 146, 167, 179